CANADA
DEPARTMENT OF MINES
Hon. Louis Coderre, Minister; R. W. Brock, Deputy Minister.

MINES BRANCH
Eugene Haanel, Ph.D., Director.

RESEARCHES ON COBALT AND COBALT ALLOYS, CONDUCTED AT QUEENS UNIVERSITY, KINGSTON, ONTARIO, FOR THE MINES BRANCH OF THE DEPARTMENT OF MINES

PART I

THE PHYSICAL PROPERTIES

OF

THE METAL COBALT

BY

H. T. Kalmus, B.Sc., Ph.D.

AND

C. Harper, B.A.

Dr. Eugene Haanel,
  Director of Mines Branch,
    Department of Mines,
      Ottawa.

Sir,—

I beg herewith to submit a report entitled, "The Physical Properties of the Metal Cobalt", the same being the second completed part of the series of investigations on cobalt and cobalt alloys, for the purpose of increasing their economic importance, which have been the subject of the special researches conducted under my direction at Queens University, Kingston, Ontario, for the Mines Branch of the Department of Mines, Ottawa.

I have the honour to be, Sir,
Your obedient servant,
Herbert T. Kalmus.

CONTENTS

PAGE

Compressive strength measurements... 25
   Test bars... 25
   Method of measurement.. 25
   "Commercial cobalt".. 25
      Cast and unannealed.. 25
      Cast and annealed.. 26
   Pure cobalt... 27
      Cast and unannealed.. 27
      Cast and annealed.. 27
   Conclusions... 28
     Pure cobalt... 28
     "Commercial cobalt".. 28
Machining, rolling and swaging of metallic cobalt.............................. 29
   Turning properties.. 29
   Swaging of cobalt... 29
     Swaging machines: description of...................................... 29
   Conclusions... 30
Measurement of electrical resistance... 31
   Description of apparatus.. 31
   Method of computation... 33
   Annealing of wires.. 33
   "Commercial cobalt".. 34
      Unannealed... 34
      Annealed... 36
   Pure cobalt... 38
      Unannealed... 38
      Annealed... 39
   Specific electrical resistance of cobalt.................................. 41
   Specific electrical resistance of nickel.................................. 41
   Conclusions:—
     Pure cobalt... 41
     "Commercial cobalt".. 42
   Magnetic permeability... 42
   Specific heat measurements.. 42
     Method and apparatus.. 42
     Material.. 43
Microphotographs... 44
   Pure cobalt H 212:   Analysis of...................................... 44
   Commercial cobalt H 213: "   "....................................... 45
   Cobalt H 214:   "   "...................................... 45
   Commercial cobalt H 130: "   "....................................... 45
      "   "   H 214c (Plate VIII): Analysis of.................... 46
      "   "   H 214c ("   IX):   "   "....................... 46
      "   "   H 87c ("   X):   "   "....................... 46
      "   "   H 87c ("   XI):   "   "....................... 47
      "   "   H 109: Analysis of............................... 47
      "   "   H 211   "   "........................... 47
Pure Nickel.. 48
Acknowledgments.. 48

ILLUSTRATIONS

Photographs.

Drawings.

RESEARCHES ON COBALT AND COBALT ALLOYS, CONDUCTED
AT QUEENS UNIVERSITY, KINGSTON, ONT., FOR THE MINES
BRANCH OF THE DEPARTMENT OF MINES

PART I

THE PHYSICAL PROPERTIES OF
THE METAL COBALT

BY

Herbert T. Kalmus, B.Sc., Ph.D.

and

C. Harper, B.A.

PART I

THE PHYSICAL PROPERTIES OF THE METAL COBALT

INTRODUCTORY.

An extended investigation of the metal cobalt, and its alloys, for the purpose of increasing its industrial and economic importance, has been, and is being conducted at the School of Mining, Queens University, Kingston, Ont., for the Mines Branch of the Department of Mines, Ottawa. The following are the principal subdivisions of the work:—

    I. THE PREPARATION OF METALLIC COBALT BY REDUCTION OF THE OXIDE.
    II. A STUDY OF THE PHYSICAL PROPERTIES OF THE METAL COBALT.
    III. ELECTRO-PLATING WITH COBALT AND ITS ALLOYS.
    IV. COBALT ALLOYS OF EXTREME HARDNESS.
    V. COBALT ALLOYS WITH NON-CORROSIVE PROPERTIES.
    VI. SPECIAL COBALT ALLOYS.

This paper is Part II of the above, and is a report on a large number of measurements made at the University laboratories, of some of the important physical and mechanical properties of metallic cobalt. The properties whic have been particularly studied are:—

    (a) Density.
    (b) Hardness.
    (c) Melting point.
    (d) Tensile breaking strength.
    (e) Tensile yield point.
    (f) Compressive breaking strength.
    (g) Compressive yield point.
    (h) Rolling and turning properties.
    (i) Electrical resistance.
    (j) Magnetic permeability.
    (k) Specific heat.

In connexion with these, a number of microphotographs have been taken.

As far as possible existing data, as found in the literature on the subject, will be reported under the above headings; but throughout, references will be cited for all facts and figures taken from the literature, so that there will be no ambiguity as to what is old and what is new.

PREPARATION OF METALLIC COBALT FOR THE STUDY OF ITS PHYSICAL PROPERTIES.

It is true of cobalt as of most metals, that its physical properties are often greatly influenced by the presence of small percentages of impurities. It is well known, for example, that less than 0·01 per cent of arsenic in copper is sufficient to account for a drop in its electrical conductivity of 3·3 per cent.[1] Similarly, for cobalt we find that a few tenths of one per cent of impurities often doubles or trebles its electrical resistance[2].

The cobalt for these investigations has been prepared by reduction of cobalt oxide, Co_3O_4. Commercial oxide was obtained from the smelters, and after a crude purification, has been reduced to form what we shall call "commercial cobalt"; again this commercial oxide has been purified to a high degree of purity, from which has been prepared what we style "pure cobalt". These two names are used in this paper largely for brevity and convenience; the analysis of each sample is given with the data of its properties. The properties of each of these have been measured, and will be discussed separately.

Preparation of "Commercial" Metallic Cobalt.

Black cobalt oxide, Co_3O_4, as obtained from the smelters, was given a crude purification, and then reduced with carbon to metallic cobalt.

This purification was carried out in accordance with the procedure given below, under "Purification of Commercial Cobalt Oxide"; except that no particular care was taken that the separations should be complete. That is to say, the purification was such as could be effected commercially at comparatively low cost, and is such as is at present attained in the manufacture of the best commercial cobalt oxide.

Three samples of cobalt oxide from which this commercial cobalt was made, analysed as follows:—

June, 1912.   %
    Co..71·99
    Fe...0·11
    Ni..0·040
    S...0·020
    Ca..0·021
    Si..0·090

November, 1912.   %
    Co..71·52
    Fe...0·27
    Ni..0·020
    Si..0·090
    S..trace
    Ca...trace

April, 1913.   %
    Co..70·4
    Fe..0·102
    Ni...trace
    Ca..0·15
    S...0·21
    Si..0·11

Analyses of the "commercial" metal produced from this oxide will be given in connexion with the measurements of its physical and mechanical properties.

---

[1] J. H. Dellinger, "The Temperature Coefficient of Resistance of Copper", Bulletin of the U. S. Bureau of Standards, Vol. 7, 1911, page 79.
[2] This paper, pp. 30-40.

3

REDUCTION WITH CARBON.

Finely divided cobalt oxide was intimately mixed with powdered charcoal, the amount of charcoal being about 10 per cent in excess of that required by the equation,

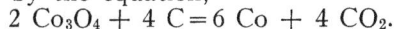

$$2\ Co_3O_4 + 4\ C = 6\ Co + 4\ CO_2.$$

This charge, usually 5 lbs. of oxide, was reduced in a graphite crucible, either with a Steele-Harvey oil fired furnace, or in a modified Hoskins electric resistor furnace. The charge was frequently stirred during the reduction and the temperature was controlled at about 1,000°C, making temperature observations with a suitable pyrometer.[1]

When the reduction was complete, or nearly so, the temperature of the furnace was raised to about 1,550°C, sufficient to melt the cobalt. Bars of cobalt were not cast at this temperature—1,550°C; for in that event the castings would be found to be permeated with blow holes. In order to obviate this, the common practice of "soaking" was adopted. In the case of cobalt this consisted in lowering the temperature of the melt from approximately 1,550°C to approximately 1,493°C, maintaining it at that temperature for about one-half hour, and then casting from this reduced temperature, which was but slightly above the melting point of the metal. The casting was into sand, or iron moulds, in forms suitable to be swaged into wires, rolled into sheets, machined into test bars, or to be used for whatever experiments were required.

The "commercial cobalt" obtained in this way, invariably contained in the neighbourhood of 0·20 per cent carbon, together with such small percentages of iron, nickel, sulphur, calcium, and silicon, as were contained in the original oxide, and not slagged off.

The metal produced in this way varied in quality from time to time, and its analysis will be given for each of the tests to be described in this paper.

METALLIC COBALT OF EXTREME PURITY.

Part of these experiments were designed to obtain measurements of the physical properties of cobalt of much more extreme purity than would ordinarily be found in industrial practice. For this purpose metal was obtained from a specially purified cobalt oxide, either by reduction of the oxide with hydrogen gas, or with carbon monoxide gas.

Purification of Commercial Cobalt Oxide.

Cobalt oxide as obtained from the smelters, and as sold on the market, usually contains small percentages of the following impurities, Ca, S, As, Si, Fe and Ni.

This commercial oxide was purified by either one of the following methods[2]:—

METHOD A.

Starting with a crude cobalt oxide, the impurities may be reduced, as far as is desired, by the following procedure.

---

[1] For a full description of experiments on the preparation of metallic cobalt by reduction of the oxide with carbon, see "Preparation of Metallic Cobalt by Reduction of the Oxide", Herbert T. Kalmus, Bulletin No. 259, Mines Branch, Canada Department of Mines, page 4, 1913.

[2] For details of the application of this method, see "Preparation of Metallic Cobalt by Reduction of the Oxide" Herbert T. Kalmus, Bulletin No. 259, Mines Branch, Canada Department of Mines, page 3, 1913.

Silica.—Dissolve the crude oxide in hydrochloric acid, according to the reaction,

$$Co_3O_4 + 8\ HCl = 3\ CoCl_2 + 4\ H_2O + Cl_2.$$

This may best be done by heating and agitating with steam. If silica is present, it will not dissolve, and may be removed by filtration or decantation. The same is true of silicates which are not decomposed by this treatment. Decomposable silicates would send a certain amount of silica into solution, which would be thrown out again during the next step, and be filtered off with the arsenic and iron.

Iron and Arsenic. To the cobalt chloride solution formed by dissolving the oxide in hydrochloric acid as above, gradually add finely divided $CaCO_3$ or pure marble, until no further precipitate is formed. This calcium carbonate addition will precipitate a heavy brown mud, which contains the iron and arsenic content of the original oxide.

Nickel.—For most purposes it will not be necessary to separate the small amount of nickel from the cobalt, but if this is desirable it may be done as follows. The cobalt chloride solution, containing a certain amount of nickel chloride, is of an intense red or claret colour. Add a solution of bleach to this mixed chloride solution until it has almost completely lost its colour. The bleach solution differentially precipitates hydrates of nickel and cobalt, so that the nickel is not appreciably brought down until the cobalt has been almost entirely precipitated.

The bleach will precipitate a black hydrated oxide of cobalt, and the diminishing redness of the solution will indicate the end point. If all of the steps above outlined have been applied to the original oxide, this final black precipitate may be calcined at about 750°C, to yield black Co_3O_4.

Sulphur.—Any sulphur which was present in the original oxide, and which has been carried through to the final product, or which may have been introduced with the bleach, may readily be removed by boiling the final dried oxide with sodium carbonate and dilute hydrochloric acid, as follows. The Na_2CO_3 reacts with the sulphur, which is in the form of sulphate after the calcination, according to the reaction,

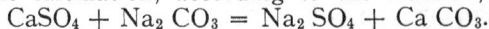

$$CaSO_4 + Na_2CO_3 = Na_2SO_4 + Ca\ CO_3.$$

The soluble sodium sulphate formed is washed out with water. A further washing is given with dilute hydrochloric acid, which decomposes the calcium carbonate into soluble calcium chloride and CO_2 gas. The $CaCl_2$ is washed out with water. This method is, of course, only applicable for the removal of small percentages of Ca and S, as they are found in the oxides in question.

METHOD *B*.

(a) Dissolve commercial Co_3O_4 in hydrochloric acid, and filter or decant off insoluble silica and silicates.

(b) Add finely divided $CaCO_3$, precipitating a heavy brown mud which contains the iron and arsenic content of the original oxide.

(c) Add barium chloride, before filtering, to precipitate sulphates as barium sulphate, and then decant and filter complete precipitate from (b) and (c).

(d) There will now remain a solution of mixed $CoCl_2$ and $NiCl_2$. To this solution add a filtered bleach[1] solution, which will differentially precipitate the nickel and cobalt as hydrates. That is, the bleach will precipitate a black hydrated oxide of cobalt, and the diminishing redness

---

[1] This bleach solution should be free from sulphates; if not, it should be rendered so by adding $BaCl_2$ and filtering.

of the solution will indicate the end point at which nickel hydrate will begin to precipitate.

This final black precipitate was dried and calcined at about 760°C yielding black Co_3O_4. By this method oxide containing only traces of impurities was obtained.

Cobalt of extreme purity, for our tests, was made from such oxide as this, by reduction with hydrogen gas or with carbon monoxide gas.

The analyses of this pure metal will be given with the various tests of its properties, the following being a typical example:—

	%
Co.	99·9
Ni.	none
Fe.	0·10
S.	0·019
Ca.	none
Si.	0·040
C.	none

REDUCTION OF Co_3O_4 WITH HYDROGEN AND CARBON MONOXIDE GAS.

The finely divided cobalt oxide was placed within an electric resistor furnace, in an alundum or silica muffle, through which a steam of hydrogen or carbon monoxide gas was passed. The temperature was controlled at about 950°C, by suitable pyrometer observations, and the excess of gas was burned at the discharging end of the furnace.[1]

The metal formed by reduction in this way, was in the form of a finely divided grey powder, if the reduction was made in the neighbourhood of 900°C, and in the form of a more or less sintered grey powder, if the reduction was made in the neighbourhood of 1000°C. It was placed in an alundum lined crucible, within an electric resistor furnace, melted, and cast in the desired form to be rolled, swaged, or turned.

The cobalt of extreme purity, prepared in this way, varied slightly in quality, and chemical analyses will be given throughout this paper accompanying the report of each of the tests.

COLOUR.

Pure metallic cobalt very much resembles nickel in colour, although when plated and polished under proper conditions, while beautifully white, it possesses a slightly bluish cast. Sometimes it deposits as a black matte. Metallic cobalt which has been reduced from the oxide at a sufficiently low temperature is a grey powder.

DENSITY—SPECIFIC GRAVITY.

The density of both cast and rolled cobalt was determined in this laboratory by the Archimedes method. A sphere or cylinder of the material was weighed, using a delicate balance, both in air and under water. The weighings were corrected in the computations for the buoyancy of the air, and the measurements reduced so as to be expressed in terms of water at its maximum density, that is, at 4°C. Following are the results:

---

[1] For a description of the method of purification of the gas, of the details of apparatus, and of the method of conducting reduction runs, see "Purification of Metallic Cobalt by Reduction of the Oxide", Herbert T. Kalmus, Bulletin No. 259, Mines Branch, Canada Department of Mines, pp. 11–31, 1913.

Density of " Commercial Cobalt."

Cast and Unannealed.

Sample number.	Date.	Form of sample.	Heat and mechanical treatment.	Analysis.		Density value at °C.
H 109	Dec. 8, 1913.	Cylindrical bar	Cast from just above melting point in iron mould, allowed to cool, and turned in lathe.	Co Ni Fe S C P	96·8 0·56 2·36 0·022 0·063 0·017	8·7997　　18·5°C. See microphotograph, Plate XII
H 130	Feb. 1914. Average of five determinations.	Thin cylindrical bar.	Cast from just above melting point in iron mould. allowed to cool in iron mould, and turned in lathe.	Co Ni Fe C S P	96·5 2·0 1·27 0·305 0·054 0·015	8·7690　　17°C See microphotograph, Plate VII.
H 87 c	Feb. 1914. Average of four determinations.	Cylindrical bar.	Cast from just above melting point in iron mould, allowed to cool in iron mould, and turned in lathe.	Co Ni Fe S C Ca Si P	97·8 0·50 1·46 0·020 0·18 trace 0·020 trace.	8·6658　　17°C. See microphotograph, Plate XI.

Density of Pure Cobalt.

Cast and Unannealed.

Sample number.	Date.	Form of sample.	Heat and mechanical treatment.	Analysis.		Density value at °C.
H 212	Jan.-Feb. 1914. Average of six determinations.	Cylindrical bar.	Cast from just above melting point, allowed to cool in iron mould, and turned in lathe.	Co Fe Ni C S Ca Si	99·9 0·20 none none 0·017 none none	8·7562　　17°C. See microphotograph, Plate IV.
H 214	Apl. 3, 1914.	Swaged wire of diameter 0·0901 cms.	Cast from just above melting point, allowed to cool in iron mould, and given special heat and mechanical treatment, according to page 30, after which it was swaged to wire.	Co Ni Fe Si Ca S C P	98·71 none 1·15 0·14 none 0·012 0·039 0·010	8·8490　　15°C. See microphotograph, Plate VI.
H 193	Nov. 15, 1913	Cylindrical bar.	Cast from just above melting point, allowed to cool in iron mould, and turned in lathe.	Co Ni Fe S Ca	99·1 none 0·80 0·021 trace	8·7889　　20°C.
H 213	Jan.-Feb. 1914. Average of three determinations.	Cylindrical bar.	Cast from just above melting point, allowed to cool in iron mould, and turned in lathe. *Cast and Annealed*	Co Fe S Ni C Si	99·73 0·14 0·019 none none 0·020	8·7732　　16°C. See microphotograph, Plate V.
H 212	Jan. 12, 1914.	Thin cylindrical bar.	Cast from just above melting point, allowed to cool in iron mould, and turned in lathe. Annealed from 700°C.	Co Fe Ni C S Ca Si	99·9 0·20 none none 0·017 none none	8·8105　　14·5°C.

PLATE I.

Brinell Hardness Testing Machine.

Rolled.

Sample number.	Date.	Form of sample.	Heat and mechanical treatment.	Analysis.		Density value at °C.	
H 213	Jan. 19, 1914.	Thin cylindrical bar.	Cast from just above melting point, cooled in iron mould, and swaged down to thin cylindrical bar.	Co Fe S Ni C Si	99·73 0·14 0·019 none none 0·020	8·9278	14°C.
H 212	Jan. 23, 1914.	Wire, 0·0840cms. diameter.	Cast from just above melting point, cooled in iron mould, and swaged down to thin cylindrical bar.	Co Fe Ni C S Ca Si	99·9 0·20 none none 0·017 none none	8·9227	19°C.

A number of density determinations of metallic cobalt, as made by other investigators, are recorded in the literature, most of which, however, were made at an early date, and very little is said of the nature of the metal. The following table of values is taken from the more recent and probably more accurate of them.

Density of Cobalt.

Tilden[1]....................................8·718   21°C

G. Neumann and F. Streintz[2]................8·6   —

Copaux[3]...................................8·8   15°C

Winkler[4] ....7·9678   —

Kalmus and Harper, unannealed..............8·7918   17·0°C

Kalmus and Harper, annealed................8·8105   14·5°C

Kalmus and Harper, swaged..................8·9253   16·5°C.

The values from the literature are generally lower than those measured by us, no doubt because of impurities in the metal, or because of the difficulties of casting without occluding a certain amount of gas.

HARDNESS.

Hardness Testing Machine.

Hardness measurements were made in this laboratory on a standard Olsen hardness testing machine of 100,000 lbs. capacity, (Tinius Olsen Co., Philadelphia, Pa.), as shown in Plate I. The machine consists of a framework on which is mounted a lever system. To one end of this lever system a penetrating ball (**A**) is attached, while at the other, weights are attached, which, when applied, cause motion of the lever system and penetration of the ball into the metal (**B**) to be tested. An instrument (**C**) is mounted on the main lever which measures automatically the actual penetration of the steel ball to 0·0001 of an inch.

Figure 1 shows the details of the essential parts of the machine. The test piece is placed on the head(**R**), which is brought into contact with the penetrating sphere by means of a screw (**S**). The sphere is one centimeter

[1] Chemical News, Vol. 78, p. 16, 1898.
[2] Monatshefte für Chemie, Vienna, Vol. 12, 1891, p. 642.
[3] Annalen de Chemie et de Physique, (8), Vol, 6, 1905, p. 508.
[4] Berg und hüttenmännische Zeitung, Vol 39, 1880, p. 87.

in diameter. A small initial pressure is applied to the piston (P). The zero reading is then taken. The desired pressure is then applied and released to the point of initial pressure before the final reading is taken. These readings are made to 0·0001 of an inch. The difference between the initial and final readings is equal to the depth of indentation of the sphere.

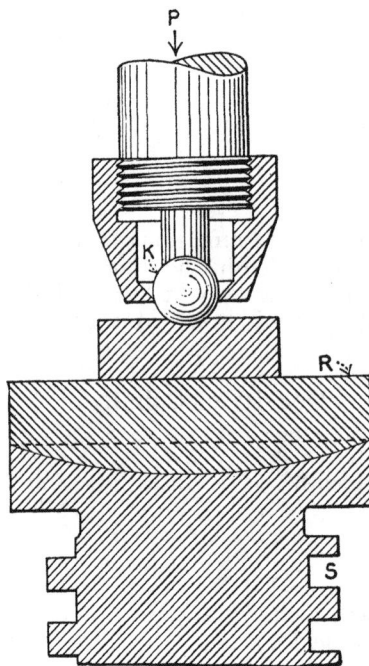

Fig. 1. Detail of Brinell hardness testing machine.

Discussion of Brinell Method of Hardness Measurement.

After the Brinell method was introduced it was shown by Benedicks,[1] Le Chatelier,[2] Leon,[3] Malmstrom,[4] Meyers[5] and others, that the Brinell hardness numeral is dependent upon the size of the sphere used to make the indentation, and upon the pressure to which it is subjected. Consequently these must be specified with each measurement.

All hardness measurements of the metal cobalt, made by us, have been computed by the Brinell system, and have been made with a sphere of one centimeter diameter, and with a load of 3500 pounds, unless stated otherwise.

We have measured the Brinell hardness of a series of common substances under the identical conditions that we have used for cobalt, and a table of these values is given below for comparison. In each instance the value is the mean of a number of observations, and they are reproducible, on the same sample, to within a few per cent. Different samples of most of this material give values differing considerably among themselves. This table is given merely to serve as a rough basis of comparison.

[1] Recherches Physiques sur l'acieur au carbone. Upsala, 1904.
[2] Revue de Metallurgie, 1906, No. 2.
[3] Die Brinellsche Harteprobe und ihre praktische Verwendung.   Proceedings, International Association for Testing Materials, 1906.
[4] Stahl und Eisen, 1907, No. 50.
[5] Untersuchungen uber Harteprufung und Härte, Zeitschrift des Vereins Deutscher Ingenieure, 1907.

Reference Table of Brinell Hardness.

Material.	Date.	Load.	Brinell hardness.
Copper, rolled sheet, unannealed,................	Jan. 1913	1000 lbs.	65·6
Copper, rolled sheet, unannealed,................	Jan. 1914	1000 lbs.	67·4
Copper, rolled sheet, unannealed,................	Jan. 1914	3500 lbs.	75·0
Copper, rolled sheet, unannealed,................	Jan. 1914	3500 lbs.	81·9
Swedish iron.................................	Jan. 1913	3500 lbs.	90·7
Swedish iron.................................	Jan. 1914	1000 lbs.	68·6
Swedish iron.................................	Jan. 1914	3500 lbs.	75·2
Wrought iron................................	Jan. 1913	3500 lbs.	92·0
Wrought iron................................	Jan. 1914	1000 lbs.	83·1
Wrought iron................................	Jan. 1914	3500 lbs.	100·2
Cast iron....................................	Jan. 1913	3500 lbs.	97·8
Cast iron....................................	Jan. 1914	1000 lbs.	84·4
Cast iron	Jan. 1914	3500 lbs.	104·5
Mild steel...................................	Jan. 1913	3500 lbs.	109·9
Mild steel, cold rolled shafting................	Jan. 1914	3500 lbs.	126·2
Tool steel...................................	Jan. 1913	3500 lbs.	153·8
Tool steel "Crescent".........................	Jan. 1914	3500 lbs.	130·2
Spring steel.................................	Jan. 1913	3500 lbs.	160·3
Spring steel.................................	Jan. 1914	3500 lbs.	178·0
Tool steel, self-hardening,.....................	Jan. 1913	3500 lbs.	180·0
Tool steel, self-hardening, "Rex" (before hardening)	Jan. 1914	3500 lbs.	162·1
Tool steel, self-hardening. "Rex", (after hardening)	Jan. 1914	3500 lbs.	240·0
Tool steel, self-hardening, from workshop (School of Mining)................................	Jan. 1914	3500 lbs.	259·0

Cobalt for Brinell Hardness Measurements.

The hardness of cobalt, like that of most other metals, is determined to a greater extent by its physical and mechanical treatment than by slight variations in its chemical composition, if we except perhaps the presence of carbon. Even our "commercial cobalt" contains but small percentages of total impurities, of which the greater part is iron and nickel, and which in the small amounts present would not greatly effect the hardness. In the samples under "commercial cobalt" the percentage of carbon is given throughout, and the percentage of other impurities is between the following limits:—

	%		%
S	0·010	to	0·070
Ca	trace	to	0·015
Si	trace	to	0·20
Fe	0·10	to	1·0
Ni	trace	to	0·50
C	0·10	to	0·60

The total impurities in any one sample of this "commercial cobalt" rarely exceeded 1·5 per cent.

Brinell Hardness Measurement.

A single measurement of the Brinell hardness is given in full to show the concordance of observations among themselves, and the details of computation. This may be taken as typical of the large number of measurements which were made.

Sample H 193,   Dec. 9, 1913.   Load 3500 lbs.

Initial Reading.	Reading under load.	Indentation in inches.	
0·0344	0·0489	0·0145	
0·0333	0·0480	0·0147	
0·0309	0·0465	0·0156	
0·0321	0·0468	0·0147	
		Average indentation	0·0149 inches== 0·379 millimeters.

Brinell Hardness by definition $= \dfrac{\text{total pressure in kilograms}}{\text{area of depression in square millimeters}} =$

$\dfrac{P}{2\,\pi\,rh}$ where

P = load in kilograms

r = radius of indenting ball in millimeters

h = depth of depression in millimeters.

Therefore, B.H. $= \dfrac{3500}{2\cdot2} \times \dfrac{1}{2 \times \pi \times 5 \times 0\cdot379} = 133\cdot4$

Measurement of Brinell Hardness of "Commercial Cobalt."

Load 3500 lbs., unless stated to the contrary.

Sample number.	Date.	Heat and mechanical treatment.	Carbon, sulphur and phosphorus content.	Brinell hardness.	Remarks.
H 109	Dec. 9, 1913.	Cast from just above melting point, allowed to cool in iron mould, and turned in lathe.	C 0·062 S 0·022 P 0·017	111·4	See microphotograph, Plate XII.
H 109	Jan. 14, 1914.	Cast from just above melting point, allowed to cool in iron mould, and turned in lathe.	C 0·062 S 0·022 P 0·017	100·9	See microphotograph, Plate XII.
H 109	Dec. 11, 1913.	Cast from just above melting point, allowed to cool in iron mould, and turned in lathe.	C 0·062 S 0·022 P 0·017	104·4	See microphotograph, Plate XII.
H 109	Dec. 22, 1913.	Cast from just above melting point, allowed to cool in iron mould, and turned in lathe.	C 0·062 S 0·022 P 0·017	111·7	Metal soft, tough and turns with medium long curling chip. See microphotograph, Plate XII.
H 109	Sept. 15, 1914.	Cast from just above melting point, allowed to cool in iron mould, and turned in lathe.	C 0·062 S 0·022 P 0·017	100·2	Metal soft, tough and turns with medium long curling chip. See microphotograph, Plate XII.
H 211	Jan. 14, 1914.	Cast from just above melting point, allowed to cool in iron mould, and turned in lathe.	C 0·18 S 0·080 P 0·031	128·2	Metal soft and medium tough. Machines with medium long curling chip. See microphotograph, Plate XIII.
H 211	Sept. 15, 1914.	Cast from just above melting point, allowed to cool in iron mould, and turned in lathe.	C 0·18 S 0·080 P 0·031	130·7	Metal soft and medium tough. Machines with medium long curling chip. See microphotograph, Plate XIII.
H 87c	Jan. 16, 1914.	Cast from just above melting point, allowed to cool in iron mould, and turned in lathe.	C 0·18 S 0·022 P 0·012	131	Metal medium hard and tough. Machines with curling chip. See microphotograph, Plate XI.
H 214c	Sept. 15, 1914.	Cast from just above melting point, allowed to cool in iron mould and turned in lathe, and annealed from 1000°.	C 0·067 S 0·012 P 0·010	136·9	See microphotograph, Plate IX.
H 214c	Sept. 15, 1914.	Cast from just above melting point, allowed to cool in iron mould and turned in lathe, and annealed from 850°.	C 0·067 S 0·012 P 0·010	138·6	See microphotograph, Plate VIII.

Sample number.	Date.	Heat and mechanical treatment.	Carbon sulphur and phosphorus content.	Brinell hardness.	Remarks.
H 214c	Sept. 15, 1914.	Cast from just above melting point, allowed to cool in iron mould and turned in lathe, unannealed	C 0·067 S 0·012 P 0·010	123·9	
H 87 a and e	Dec. 22, 1913.	Cast from just above melting point, allowed to cool in iron mould, and turned in lathe.	C 0·22 S 0·030 P none	119·2	
H 87 a and e	Sept. 15, 1914.	Cast from just above melting point, allowed to cool in iron mould, and turned in lathe.	C 0·22 S 0·030 P none	132·9	
H 130	Dec. 11, 1913.	Cast from just above melting point, allowed to cool in iron mould, and turned in lathe.	C 0·305 S 0·054 P 0·015	115	Metal short grained, brittle, and turns with short chip.
H 130	Sept. 15, 1914.	Cast from just above melting point, allowed to cool in iron mould, and turned in lathe.	C 0·305 S 0·054 P 0·015	113·8	
H 130	Jan. 14, 1914.	Cast from just above melting point, allowed to cool in iron mould, and turned in lathe.	C 0·305 S 0·054 P 0·015	116·6	Metal soft and tough. Machines with curling chip. See microphotograph, Plate VII.
H 87 d and b	Feb. 1, 1913.	Cast from just above melting point, allowed to cool in iron mould, and turned in lathe.	C 0·36 S 0·016 P none	112	Very tough to turn in lathe.
H 87 b and d	Feb. 1, 1913.	Cast from just above melting point, allowed to cool in iron mould, and turned in lathe.	C 0·37 S 0·015 P none	117	Very tough to turn in lathe.

Measurement of Brinell Hardness of Pure Cobalt.
Load 3500 lbs., unless stated to the contrary.

Sample number.	Date.	Heat and mechanical treatment.	Analysis.	Brinell hardness.	Remarks.
H 193	Nov. 15, 1913	Cast from just above melting point, allowed to cool in iron mould, and turned in lathe.	Co 99·10 Ni none Fe 0·80 S 0·020 C none Ca trace	129·7	Load 2,500 lbs. Metal soft and brittle. Machines with short chip.
H 193	Dec. 9, 1913	Same	Same	133·4	
H 193	Nov. 15, 1913	Same	Same	131·2	
H 212	Jan. 9, 1914	Same	Co 99·9 Ni none Fe 0·20 S 0·017 C none	105·5	Load 2,500 lbs.
H 212	Jan. 14, 1914	Same	Same	128·7	Metal soft and brittle. Machines with short chip. See microphotograph, Plate IV.
H 212	Jan. 14, 1914	Cast from just above melting point, allowed to cool in iron mould, turned in lathe, and annealed from 700° C.	Co 99·9 Ni none Fe 0·20 S 0·017 C none	130·8	Load 2,500 lbs.
H 213	Jan. 14, 1914	Same	Co 99·73 Ni none Fe 0·14 S 0·019 C none	121	Metal soft and brittle. Machines with short chip. See microphotograph, Plate V.
H 217	Sept. 15, 1914	Annealed two hours at 600° C., allowed to cool slowly, and again machined.	Co 99·55	125·9	
H 217	Sept. 15, 1914	Cast from just above melting point, allowed to cool in iron mould, and machined.	Same	109·5	

Hardness of Cobalt as Observed by Other Investigators.

There is very little in the literature on the hardness of cobalt, except a few more or less qualitative observations. However, a careful measurement seems to have been made by R. Reur and K. Kaneko[1], from which they compute the Brinell hardness of cobalt to be 132.

Comparative Hardness of Nickel and Cobalt.

For comparison we have measured the hardness of both cast and sheet nickel under the same conditions that we have used for cobalt, load 3500 lbs., and found them to be respectively 83·1 and 85·1 Brinell, the latter for a $\frac{1}{4}$ inch sheet. An independent check test on the hardness of cast nickel, gave as a result 76·4.

From these measurements it is therefore apparent that the hardness of cobalt is considerably greater than that of either iron or nickel, under corresponding conditions.

Conclusions.

(1). The above tables show the Brinell hardness of cobalt cast from just above the melting point, and allowed to cool in an iron mould, to be in the neighbourhood of 124·0 (load 3500 lbs.). This is the mean of nine observations with an average deviation from the mean, of 7·9.

(2). The hardness of cobalt, cast from just above its melting point, is considerably higher than that of cast iron or cast nickel under corresponding conditions.

(3). The effect of the addition of 0·060 to 0·37 per cent of carbon on the hardness of "commercial" cobalt is not sufficient to offset the effect of slight variations in heat treatment. The measurements are not sufficiently concordant to warrant drawing general conclusions.

MELTING TEMPERATURE OF COBALT.

A considerable number of melting point determinations of the metal cobalt were made in an Arsem electric vacuum furnace, (General Electric Company, Schenectady, N. Y.). These determinations were made by the usual cooling or melting curve method, using pure alumina crucibles, and alundum lined graphite crucibles.

Cobalt has a very sharp melting point, differing in this respect from iron, which becomes plastic as it approaches its melting point. With iron, the actual temperature of melting is not sharply defined, there being a considerable transition region; whereas with cobalt quite the reverse is true. Therefore, the melting point of cobalt may be determined with accuracy by the cooling and melting curve method.

The Melting Furnace.

The essential feature of the furnace is the heating element, which is a spiral of graphite, through which an appropriate electric current is passed. The crucible is placed within the carbon spiral, and both are situated within a vacuum chamber. The iron casing forming the shell of the furnace, and serving as the vacuum chamber, is in constant connexion with a suitable vacuum pump. This pump was in operation during the entire melting point determination.

By this means there was no oxidizing atmosphere to shorten the life of

[1] Ferrum, Vol. 10, p. 257.
Chemical Abstracts, 1913, p. 3591.

the carbon spiral, nor any gas to carry away, by conduction, the heat developed in the spiral. A further carbon radiation shield surrounded the heating spiral, so that the temperature of this furnace may be maintained as high as 3000°C.

The melting point determinations were, therefore, made substantially in vacuum, and free from oxidizing atmospheric conditions.

Temperature Measurements.

Temperature observations were made with a Wanner optical pyrometer, which was checked before and after each set of runs against an amyl-acetate lamp standard, in accordance with a calibration certificate from the Physicalisch-Technische Reichsanstalt, at Charlottenburg. This pyrometer was also used to measure the melting points of copper and nickel during the period of its use for the determinations on cobalt, which measurements agreed with the calibration curve used, to within a few degrees. For this work the melting point of nickel was considered to be 1444°C, and of copper to be 1084°C.

The melting point of nickel, considering our calibration curve from the Reichsanstalt to be correct, was determined six times as follows:—

	Deviation from the mean.
1438°C	6
1437°C	7
1445°C	1
1446°C	2
1448°C	4
1450°C	6
1444°C mean	4·3°C average deviation of single observation from the mean.

The nickel used for these melting point measurements analysed as follows:—

	%	
Ni	99·29	
Fe	0·48	
S	0·025	See microphotograph,
Si	0·042	Plate XIV.
Ca	none	
C	none	
Co	none	
Total	99·84	

The cobalt used for these melting point determinations analysed as follows:—

	%	
Co	99·9	
Ni	none	Sample No. 212.
Fe	0·20	See microphotograph,
S	0·017	Plate IV.
Ca	none	
Si	none	
C	none	
Total	100·12	

Melting and Freezing Curves.

A cooling curve was obtained by regulating the current through the furnace, so that the charge was brought to a temperature about 100°C above its melting temperature, and then lowering the current to such a predetermined magnitude, that gradual cooling took place to about 100° below the melting temperature. During the cooling, temperature readings were made at 10, 15, or 20 second intervals.

It is obvious that a substance, if allowed to cool more or less uniformly from above its melting point, under fixed external conditions, will cease to drop uniformly in temperature when the melting point is reached, due to the latent heat of fusion which is developed within the substance upon solidifying. If we plot time as abscissae, and the temperature of the cooling mass as ordinates, in general we may get one or the other of three types of curve, corresponding with the three following typical kinds of transformation:—

(a) The substance remains at a constant temperature throughout the transformation of melting or freezing.

(b) The substance cools at a reduced rate, more or less constant during the transformation of melting or freezing.

(c) The substance undergoes an increase in temperature during the first part of the transformation.

Methods of computing the true, corrected, melting temperature from each of the above forms of curve, have been set forth in the literature by several authors discussing thermal analyses.

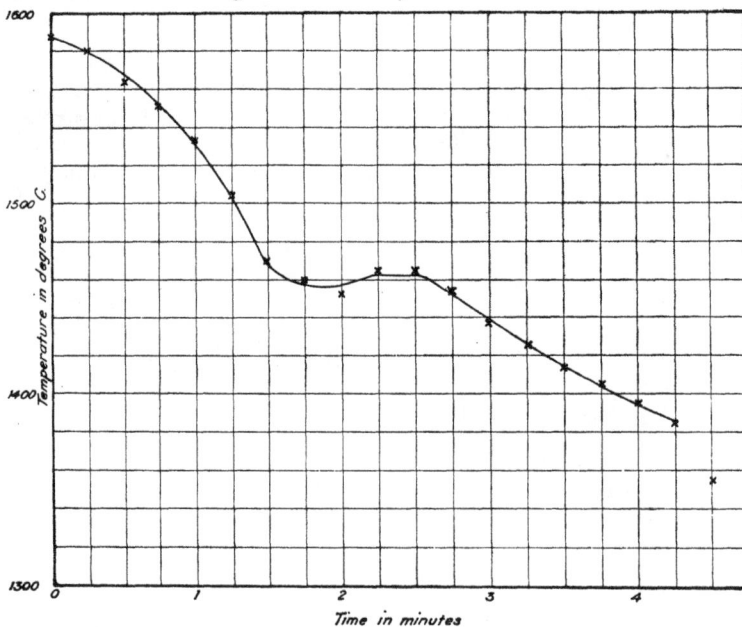

Fig. 2. Cooling curve—melting point—cobalt.

The first of the three cases indicated is by far the simplest, and may be used with accuracy for cobalt. Two sets of data, with the corresponding plot or cooling curves, taken from some fifteen of them, and which are typical of them all, are given below, to show how nearly to case (a) our melting and freezing curves correspond.

Melting Point Determination—Cobalt.

January 19, 1914. Time elapsed in minutes since beginning of run.	Readings every 15 seconds. Wanner optical pyrometer readings.	Temperature degrees C.
Run I		
See Fig. 2 0	78·2	1587
¼	77·9	1580
½	77·3	1565
¾	76·8	1553
1	75·9	1533
1¼	74·6	1505
1½	72·8	1470
1¾	72·0	1460
2	71·6	1453
2¼	72·2	1463
2½	72·2	1463
2¾	71·8	1455
3	70·6	1438
3¼	69·8	1425
3½	69·0	1415
3¾	68·5	1405
4	67·6	1395
4¼	66·8	1385
4½	64·2	1355
Run II		
See Fig. 3 0	78·9	1607
¼	77·4	1567
½	76·2	1540
¾	75·8	1530
1	74·9	1512
1¼	74·3	1500
1½	73·6	1486
1¾	72·8	1470
2	72·0	1460
2¼	72·1	1462
2½	72·0	1460
2¾	72·0	1460
3	71·8	1455
3¼	71·6	1453
3½	69·8	1425
3¾	67·4	1392
4	65·0	1364
4¼	63·2	1343
4½	62·0	1332

Without giving the details, except for the two runs above, the following is a table of the melting point determinations of cobalt analysing 99·9 per cent pure, as shown on page 13. It was sample No. 212. See Micro-photograph, Plate IV.

Summary of Results of Melting Point Determinations.

Date of run	Determined melting point of Cobalt.	Deviation of single observation from the mean
Sept. 27, 1912	1474°C	7
Sept. 30, 1912	1472	5
Sept 30, 1912	1472	5
Oct. 1, 1912	1470	3
Oct. 1, 1912	1472	5
Jan. 13, 1914	1467	0
Jan. 13, 1914	1467	0
Jan. 14, 1914	1460	7
Jan. 15, 1914	1453	14
Jan. 15, 1914	1468	1
Jan. 15, 1914	1468	1
Jan. 19, 1914	1462	5
Jan. 19, 1914	1460	7
Jan. 19, 1914	1470	3
Jan. 19, 1914	1462	5
Jan. 19, 1914	1470	3
	Mean 1467°C	Mean 4·4°C

Mean melting temperature 1467°C.   Average deviation of single observation from the mean 4·4°C.

Therefore, from these observations the melting point of pure cobalt would appear to be 1,467°C $\mp$ 1·1°C. Note, however, conclusions below.

The Melting Point of Cobalt as Determined by other Investigators.

The following values of the melting point of cobalt are taken from the literature:—

Investigator.	Melting temperature.	Purity per cent.	Method of measurement.	Reference.
Burgess and Waltenberg	1477°C	99·95	Micropyrometer	Bulletin, Bureau of Standards, Vol. 9, pp. 475. Vol. 10, pp. 13, 1913.
Burgess and Waltenberg	1478°C	99·95	Crucible melts in electric furnace	Bulletin, Bureau of Standards, Vol. 9, pp. 475. Vol. 10, pp. 13. 1913.
G. K. Burgess	1464°C	very pure	Micropyrometer	Bulletin, Bureau of Standards, Vol. 3, pp. 350.
H. Copaux	1530	not given	Interpolation between gold and platinum points.	Annalen de Chimie et de Physique (8), Vol. 6, 1905, pp. 508.
Guertler and Tamman	1528	98·3 rest largely Ni and Fe.	Cooling curve	Zeitschrift für Anorganische Chemie, Vol. 42, p. 353, 1904.
Guertler and Tamman	1468	98·3	Cooling curve	Their value above, (1528°C), corrected for melting point of nickel = 1451 instead of 1484 as taken by them.
Guertler and Tamman	1505	98·3	Cooling curve	Zeitschrift für Anorganische Chemi, Vol. 45, 1905, p. 223.
Guertler and Tamman	1455	98·3	Cooling curve	Their value above, (1505°C), corrected for melting point of nickel = 1451 instead of 1484 as taken by them.
R. Ruer and K. Kaneko	1491			Ferrum, Vol. 11, 1913, pp. 33–39.

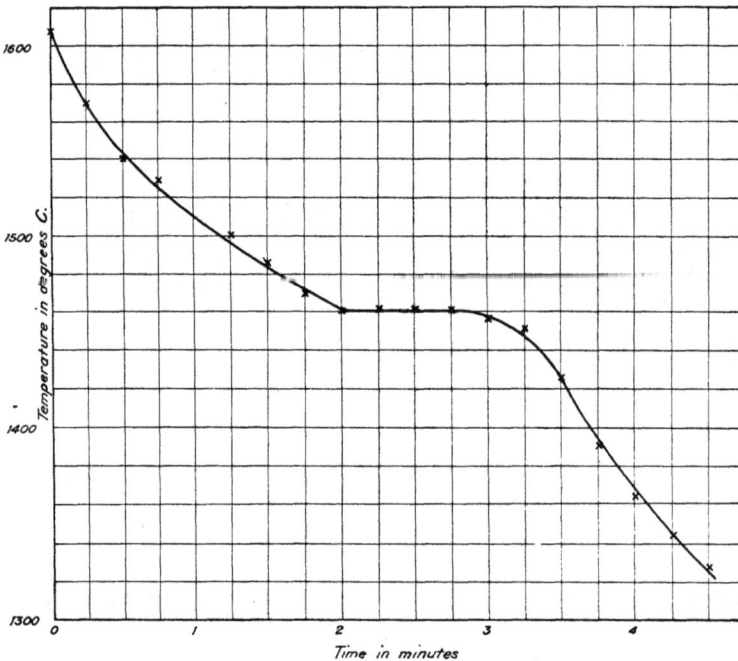

Fig. 3.   Cooling curve—melting point—cobalt.

It should be noted that Burgess and Waltenberg[1] use the value 1,452°C as the melting temperature of nickel. If we adopt this value instead of 1,444°C (see page 13) our value for the melting temperature of cobalt would be practically identical with theirs, namely, 1,478°C.

Conclusions.

The melting temperature of cobalt, as determined by us, is 1,467 $\mp$ 1·1°C. This is for metal 99·9 per cent pure, and is the mean of sixteen determinations by the cooling and heating curve method.

This value of the melting temperature is based upon pyrometer calibration curves, considering the value of the melting temperature of nickel to be 1,444°C. If we adopt the more probable value 1,452°C, for the melting temperature of nickel, **our melting temperature for cobalt is 1,478°C ± 1·1°C.**

TENSILE STRENGTH MEASUREMENTS.

Tensile Strength Testing Machine.

The tensile strength tests were made on a Riehlé Universal Standard Vertical Screw Power Testing Machine, (Riehlé Testing Machine Co., Philadelphia, Pa.), of 100,000 lbs. capacity, operated by direct connexion to an electric motor. This machine is in the testing laboratory of the Department of Civil Engineering, School of Mining, Queens University, Kingston, Ontario. The writers wish to express their thanks to Professor A. Macphail, in charge of that department, for many valuable suggestions in connexion with the use of this machine.

Test Bars.

All bars for tensile strength measurements have been "Proportional Bars," as recommended and adopted by the International Association for Testing Materials. Fig. 4 shows the shape and dimensions of these bars.

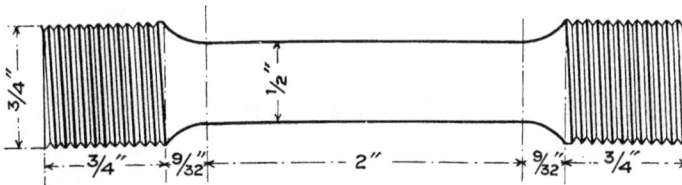

Fig. 4. Tensile strength proportional bar.

Method of Measurement.

Plate II is a picture of the testing machine used for the measurements. It was operated in the standard manner, with a chart in place, as shown at **A**, to get an autographic stress-strain diagram for each specimen.

The form of these tensile stress-strain diagrams obtained on all of our samples is shown in Fig. 5. The point **Y**, however, is often less marked than in the figure shown, and **P** and **Y** are usually very close together.

---

[1] Bulletin, United States Bureau of Standards, Vol. 10, page 6, 1913.

From the point **O** to the point **P** the elongation or strain of the material was very slight, and was proportional to the stress or load applied. **P** is the proportional limit, and **Y** is the yield point. **M** is the maximum stress, and **R** is the rupture stress or tensile breaking load, which values are given in our tables.

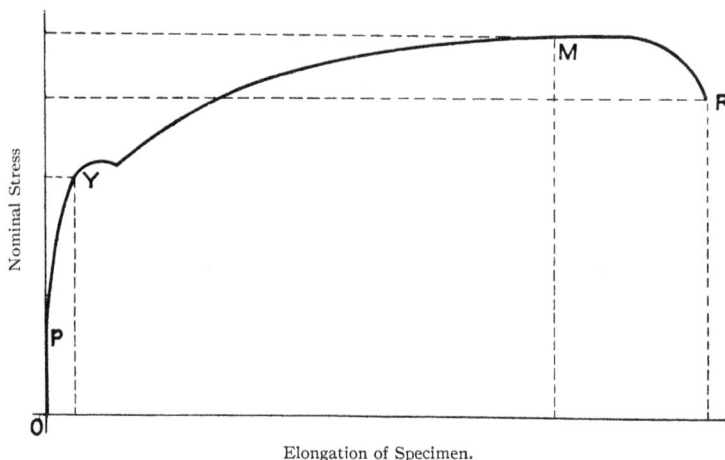

Fig. 5. **Typical tensile stress-strain diagram.**

P = Proportional limit.
Y = Yield point.
M = Maximum stress.
R = Tensile breaking load.

Below are tables of the data for these measurements as obtained by us on "commercial" cobalt and on pure cobalt.

Tensile Strength Measurements of "Commercial Cobalt."

Cast and Unannealed.

Sample number.	Date.	Heat and mechanical treatment.	Analysis.	Tensile breaking load in pounds per sq. inch.	Yield point in pounds per sq. inch.	Percentage reduction in area.	Percentage elongation in 2 inches.	Remarks.
H 109	Dec. 9, 1913.	Cast from just above melting point, allowed to cool in iron mould, and turned in lathe.	Co 96·8 Ni 0·56 Fe 2·36 S 0·022 C 0·062 P 0·017	18700	33800	7·7	3·3	Fairly fine grained fracture.
H 109	Dec. 9, 1913.	Cast from just above melting point, allowed to cool in iron mould, and turned in lathe.	Co 96·8 Ni 0·56 Fe 2·36 S 0·022 C 0·062 P 0·017	52800	33800	8·7	7·0	
H 109	Dec. 16, 1913.	Cast from just above melting point, allowed to cool in iron mould, and turned in lathe.	Co 96·8 Ni 0·56 Fe 2·36 S 0·022 C 0·062 P 0·017	57200	15300	24·5	21·9	Tough and difficult to machine in lathe. Long curling chip.
H 109	Dec. 22, 1913.	Cast from just above melting point, allowed to cool in iron mould, and turned in lathe.	Co 96·8 Ni 0·56 Fe 2·36 S 0·022 C 0·062 P 0·017	64100	9360	25·4	27·0	Fracture coarsely granular, and not uniform in appearance.

PLATE II.

Riehlé, Universal, Standard, Vertical, Screw-Power Testing Machine, 100,000 lbs. capacity.

Cast and Unannealed—(continued).

Sample number.	Date.	Heat and mechanical treatment.	Analysis.		Tensile breaking load in pounds per square inch.	Yield point in pounds per square inch.	Per-centage re-duction in area.	Per cent-age elong-ation in 2 inches.	Remarks.
H 15	Feb. 15, 1913.	Cast from just above melting point, allow-ed to cool in iron mould, and turned in lathe.	Co Fe Ni S C Ca Si P	98·5 1·0 0·30 0·020 0·18 none none none	76700	33800	7·7	6·5	
H 87c	Jan. 23, 1914.	Cast from just above melting point, allow-ed to cool in iron mould, and turned in lathe.	Co Fe Ni C S Si Ca P	97·8 1·46 0·50 0·18 0·020 0·020 none 0·012	56100	30600	5·3	5·0	Metal medium hard and tough. Mach-ines with me-dium long cur-ling chip. See microphoto-graph. Plate XI.
H 87f	Apl. 5, 1913.	Cast from just above melting point, allow-ed to cool in iron mould, and turned in lathe.	Co Fe Ni C S Si Ca P	98·7 0·80 0·20 0·22 0·029 0·020 none none	75000	33700	25·4	29·7	Very tough to machine in lathe.
H 87b	Feb. 10, 1913.	Cast from just above melting point, allow-ed to cool in iron mould, and turned in lathe.	Co Fe Ni C S Si Ca P	98·5 0·80 0·20 0·37 0·016 0·020 none none	63200	33100	24·1	24·0	
H 87d	Feb. 10, 1913.	Cast from just above melting point, allow-ed to cool in iron mould, and turned in lathe.	Co Fe Ni C S Si Ca P	98·5 0·80 0·20 0·37 0·016 0·020 none none	77700	33900	23·8	24·0	
H 211	Dec. 22, 1913.	Cast from just above melting point allow-ed to cool in iron mould, and turned in lathe.	C S P	0·18 0·080 0·031	31000	31000	none	none	Low value due to segregation of impurities. See micropho-tograph. Plate XIII.

Cast and Annealed.

Sample number.	Date.	Heat and mechanical treatment.	Analysis.		Tensile breaking load	Yield point	Per-centage re-duction	Per cent-age elong-ation	Remarks.
H 109	Jan. 9, 1914.	Cast from just above melting point, allow-ed to cool in iron mould, and turned in lathe. Annealed from 700°C.	Co Ni Fe S C Ca Si P	96·8 0·56 2·36 0·022 0·062 none none 0·017	56100	29300	13·3	13·0	Metal soft and tough. Mach-ines with med-ium long curl-ing chip.
H 109	Jan. 14, 1914.	Cast from just above melting point, allow-ed to cool in iron mould, and turned in lathe. Annealed from 700°C.	Co Ni Fe S C Ca Si P	96·8 0·56 2·36 0·022 0·062 none none 0·017	52600	31600	13·3	13·5	See microphoto-graph. Plate XII.
H 87c	Apl. 22, 1914.	Cast from just above melting point, allow-ed to cool in iron mould, and turned in lathe. Annealed at 850°C.	Co Fe Ni C S Si Ca P	97·8 1·46 0·50 0·18 0·020 0·020 none 0·012	60200	40800	·5	1·5	Very fine grain, uniform. See microphoto-graph. Plate X.

Cast and Annealed—(continued).

Sample number.	Date.	Heat and mechanical treatment.	Analysis.		Tensile breaking load in pounds per square inch.	Yield point in pounds per square inch.	Percentage reduction in area.	Percentage elongation in 2 inches.	Remarks.
H 87c	May 19, 1914	Cast from just above melting point, allowed to cool in iron mould, and turned in lathe. Annealed at 850°C.	Co Fe Ni C S Si Ca P	97·8 1·46 0·50 0·18 0·020 0·020 none 0·012	55700			2·0	
H 214C	Apl. 22, 1914.	Cast from just above melting point, allowed to cool in iron mould, and turned in lathe. Annealed at 850°C.	Co Ni Fe C S Mn Si Ca P	97·09 none 1·45 0·067 0·012 2·04 0·011 none 0·010	70500	37100	5·1	8·0	Very fine grain uniform. See microphotograph, Plate VIII.
H 214C	Apl. 22, 1914.	Cast from just above melting point, allowed to cool in iron mould, and turned in lathe. Annealed at 1000° C.	Co Ni Fe C S Mn Si Ca P	97·09 none 1·45 0·067 0·012 2·04 0·011 none 0·010	75200	25500	6·1	10·0	Very fine grain, uniform. See microphotograph, Plate IX.
H 87c	June 2, 1914.	Cast from just above the melting point, allowed to cool in iron mould, and turned in lathe. Annealed at 850°C.	Co Fe Ni C S Si Ca P	97·8 1·46 0·50 0·18 0·020 0·020 none 0·012	63,800	61,300	0·61	0·5	Very fine grain,
H 87c	June 2, 1914.	Cast from just above the melting point, allowed to cool in iron mould, and turned in lathe. Annealed at 850°C.	Co Fe Ni C S Si Ca P	97·8 1·46 0·50 0·18 0·020 0·020 none 0·012	58,000	56,100	1·5	0·3	Very fine grain.
H 87c	June 10, 1914.	Cast from just above the melting point, allowed to cool in iron mould, and turned in lathe. Annealed at 950°C.	Co Fe Ni C S Si Ca P	97·8 1·46 0·50 0·18 0·020 0·020 none 0·012	57,000	18,000	2·57	2·0	Fracture fine grained, but not uniform in appearance.
H 87c	June 10, 1914.	Cast from just above the melting point, allowed to cool in iron mould, and turned in lathe. Annealed at 950°C.	Co Fe Ni C S Si Ca P	97·8 1·46 0·50 0·18 0·020 0·020 none 0·012	58,500	20,400	3·4	2·0	Fracture fine grained, but not uniform in appearance.
H 87c	June 15, 1914.	Cast from just above the melting point, allowed to cool in iron mould, and turned in lathe. Annealed at 1000°C.	Co Fe Ni C S Si Ca P	97·8 1·46 0·50 0·18 0·020 0·020 none 0·012	65,000	55,000	3·1	1·9	Fine grained fracture.
H 87c	June 15, 1914.	Cast from just above the melting point, allowed to cool in iron mould, and turned in lathe. Annealed at 1000°C.	Co Fe Ni C S Si Ca P	97·8 1·46 0·50 0·18 0·020 0·020 none 0·012	62,300	40,800	3·1	1·9	Fine grained fracture.

Cast and Annealed—(continued).

Sample number.	Date.	Heat and mechanical treatment.	Analysis.	Tensile breaking load in pounds per sq. inch.	Yield in point pounds per sq. inch	Per-centage Reduc-tion in area.	Percent-age elonga-tion in 2 inches.	Remarks.
H 87c	June 17, 1914.	Cast from just above the melting point, allowed to cool in iron mould, and turned in lathe. Annealed at 1000°C.	Co 97·8 Fe 1·46 Ni 0·50 C 0·18 S 0·020 Si 0·020 Ca none P 0·012	42,600	40,800	0·61	0·5	Fine grained fracture.
H 87c	June 18, 1914.	Cast from just above the melting point, allowed to cool in iron mould, and turned in lathe. Annealed at 1000°C.	Co 97·8 Fe 1·46 Ni 0·50 C 0·18 S 0·020 Si 0·020 Ca none P 0·012	56,000	46,000	1·3	1·0	Fine grained fracture.

Tensile Strength Measurements of Pure Cobalt.

Cast and Unannealed.

Sample number.	Date.	Heat and mechanical treatment.	Analysis.	Tensile breaking load in pounds per square inch.	Yield point in pounds per square inch.	Per-centage re-duction in area.	Per-centage elonga-tion in 2 inches.	Remarks.
H 212	Jan. 9, 1914.	Cast from just above melting point, allow-ed to cool in iron mould, and turned in lathe	Co 99·9 Ni none Fe 0·20 S 0·017 C none	29,600	10,200	1·5	2·0	Fracture coarse grained and crystalline.
H 212	Jan. 16, 1914.	Cast from just above melting point, allow-ed to cool in iron mould, and turned in lathe.	Co 99·9 Ni none Fe 0·20 S 0·017 C none	35,400	31,400	0·5	4·0	Metal soft and brittle. Ma-chines with short chip.
H 212	Jan. 23, 1914.	Cast from just above melting point, allow-ed to cool in iron mould, and turned in lathe.	Co 99·9 Ni none Fe 0·20 S 0·017 C none	43,400	43,400	none	none	Fracture coarse grained and crystalline.
H 212	Jan. 23, 1914.	Cast from just above melting point, allow-ed to cool in iron mould, and turned in lathe.	Co 99·9 Ni none Fe 0·20 S 0·017 C none	45,800	25,500	1·5	0·5	Fracture coarse with radially crystalline structure.
H 212	Jan. 26 1914.	Cast from just above melting point, allow-ed to cool in iron mould, and turned in lathe.	Co 99·9 Ni none Fe 0·20 S 0·017 C none	23,000	23,000	none	none	Fracture coarse with radially crystalline structure. Me-tal soft and brittle. Ma-chines with short chip.
H 212	Feb. 3, 1914.	Cast from just above melting point, allow-ed to cool in iron mould, and turned in lathe.	Co 99·9 Ni none Fe 0·20 S 0·017 C none	37,900	37,900	none	none	Fracture good with fine grain. See micropho-tograph. Plate IV.
H 213	Jan. 23, 1914.	Cast from just above melting point, allow-ed to cool in iron mould, and turned in lathe.	Co 99·73 Ni none Fe 0·14 S 0·019 C none	30,600	30,600	none	none	Fracture coarse grained, cryst-alline. Metal soft and brit-tle. Machines with short chip
H 213	Jan. 23, 1914.	Cast from just above melting point, allowed to cool in iron mould, and turned in lathe.	Co 99·73 Ni none Fe 0·14 S 0·019 C none	30100	30100	none	none	Fracture coarse grained with radially crystal-line structure.

Cast and Unannealed—(continued).

Sample number.	Date.	Heat and mechanical treatment.	Analysis.	Tensile breaking load in pounds per square inch.	Yield point in pounds per square inch.	Percentage reduction in area.	Percentage elongation in 2 inches.	Remarks.
H 213	Jan. 23, 1914.	Cast from just above melting point, allowed to cool in iron mould, and turned in lathe.	Co 99·73 Ni none Fe 0·14 S 0·019 C none	23000	23000	none	none	Fracture coarse grained with radially crystalline structure. Metal soft and brittle. Machines with short chip.
H 213	Feb. 3, 1914.	Cast from just above melting point, allowed to cool in iron mould, and turned in lathe.	Co 99·73 Ni none Fe 0·14 S 0·019 C none	45300	23000	3·0	3·6	Fracture good, fine grained. See microphotograph, Plate V.

Cast and Annealed.

Sample number.	Date.	Heat and mechanical treatment.	Analysis.	Tensile breaking load in pounds per square inch.	Yield point in pounds per square inch.	Percentage reduction in area.	Percentage elongation in 2 inches.	Remarks.
H 212	Jane 14, 1914.	Cast from just above melting point, allowed to cool in iron mould, and turned in lathe. Annealed from 700°C.	Co 99·9 Ni none Fe 0·20 S 0·017 C none Si none Ca none	41200	41200	none	none	Fracture coarse grained, crystalline.
H 212	Jane 27, 1914.	Cast from just above melting point, allowed to cool in iron mould, and turned in lathe. Annealed from 700°C.	Co 99·9 Ni none Fe 0·20 S 0·017 C none Si none Ca none	28100	28100	none	none	Fracture fairly fine structure. See microphotograph, Plate IV.
H 217	June 16, 1914.	Cast from just above melting point, allowed to cool in iron mould, and turned in lathe. Annealed at 950°C.	Co 99·20 Ni none Fe 0·730 S 0·016 Ca none Si 0·091 P 0·0077 C 0·036 Al 0·021	34800	26600	0·30	0·25	Fracture fine grained and uniform.
H 217	June 17, 1914.	Cast from just above melting point, allowed to cool in iron mould, and turned in lathe. Annealed at 950°C.	Co 99·20 Ni none Fe 0·730 S 0·016 Ca none Si 0·091 P 0·0077 C 0·036 Al 0·021	43600	30600	1·3	1·0	Fracture fine grained and uniform.

Tensile Strength of Cobalt Wires.

Sample number.	Date.	Heat and mechanical treatment.	Analysis.	Tensile breaking load in pounds per square inch.	Yield point in pounds per square inch.	Percentage reduction in area.	Percentage elongation in 2 inches.	Remarks.
H 213	Jan. 29, 1914.	This sample swaged down to fine wire after special heat and mechanical treatment, which is described under "Swaging of Cobalt into Wires," pp. 29, 30.	Co 99·73 Ni none Fe 0·14 S 0·019 C none	101800		5·0	8·2	Diameter of wire 0·117 inches.
H 213	Feb. 11, 1914.	This sample swaged down to fine wire after special heat and mechanical treatment, which is described under "Swaging of Cobalt into Wires," pp. 29, 30.	Co 99·73 Ni none Fe 0·014 S 0·019 C none	77000		1·0	2·0	Diameter of wire 0.124 inches.
H 213	Mch. 24, 1914.	This sample swaged down to fine wire after special heat and mechanical treatment, which is described under "Swaging of Cobalt into Wires," pp. 29, 30.	Co 99·73 Ni none Fe 0·014 S 0·019 C none	90500		8·3		Diameter of wire 0.076 inches. Fine grained fracture.

The literature on the tensile strength of cobalt is very meagre, although Copaux[1] gives the following values for cobalt and nickel: cobalt 69,000 lbs. per square inch; nickel 58,000 lbs. per square inch.

We measured the tensile strength of pure iron and pure nickel at this laboratory, under conditions similar to those for our cobalt tensile strength measurements. The iron used for these measurements analysed:—

%
Fe.............................99·9
S..............................0·023
P..............................0·004
C..............................0·010
Mn.............................0·031
Cu.............................0·028
Si.............................trace.
Ni.............................none.
Co.............................none.

The nickel used for the tensile strength tests analysed:—

%
Ni.............................99·29
Fe.............................0·48
S..............................0·025
Si.............................0·042
Ca.............................none.
C..............................none.

A large series of measurements on nickel and iron would be required to fix the value of the tensile strength with any definiteness; our measurements show it to be approximately as follows:—

Cast nickel, tensile breaking load 18,000 lbs. per square inch.

Cast iron, tensile breaking load 23,000 lbs. per square inch.

The rate of cooling of cast metals from the fluid to the solid state is such an important factor in determining the mechanical properties of the metal, that it is just as necessary to know the dimensions of the test bars as it is to know the chemical composition. The above values for iron, nickel and cobalt, all of which have been made under exactly the same conditions, with a test bar as shown in Fig. 4, are strictly comparable, although they should not be compared with values obtained by other observers on bars of different sizes.

Conclusions.

Pure Cobalt.

I. The tensile strength of pure cobalt, cast and unannealed, is in the neighbourhood of 34,400 lbs, per square inch. This is the average of ten measurements on cobalt cast from just above its melting point, allowed to cool in iron mould, and machined in lathe to test bars.

II. The effect on the tensile strength of annealing cast cobalt, is to increase its value slightly, although this effect is not marked. The average value of our determinations was 36,980 lbs. per square inch for the annealed samples, as compared with 34,400 lbs. per square inch for the unannealed samples.

---

[1] Annalen de Chemie et de Physique, (8), Vol. 6. 1905, p. 508.

III. The percentage reduction in area and elongation are small for cast pure cobalt, as would be expected for the pure metal.

IV. The tensile yield point for pure cobalt is, in general, very close to the tensile breaking load.

V. The tensile strength of pure cobalt increases very rapidly as the metal is rolled, as is common for most metals. It may easily reach over 100,000 lbs. per square inch, by being swaged down to a wire.

VI. The tensile breaking load of pure cobalt, cast from just above the melting temperature, allowed to cool in iron mould and turned in lathe to test bar, is greater than either that of iron or nickel prepared and tested under the same conditions.

"Commercial Cobalt."

VII. The effect of the addition of carbon is to increase the tensile breaking strength of cobalt very markedly, the value rising from 34,400 lbs. per square inch for the pure cast and unannealed metal, to in the neighbourhood of 61,000 lbs. per square inch for cobalt carrying from 0·060 to 0·30 per cent carbon. More exactly, the average of eight measurements, with a carbon content of approximately 0·062 per cent, is 59,700 lbs. per square inch. Similarly, the average of fifteen measurements, with a carbon content varying in the neighbourhood of 0·25 per cent ·s 61,900 lbs. per square inch

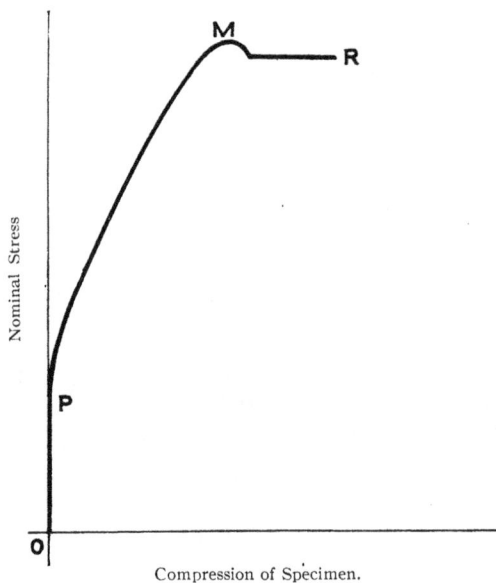

Fig. 6. **Typical compressive stress-strain diagram.**

P = Proportional limit.
M = Maximum stress.
R = Compressive breaking load.

The average deviation, however, of these individual measurements among themselves is such that, no more specific conclusion can be attached thereto. These values refer to cobalt cast from just above the melting point, allowed to cool in iron mould, machined in a lathe, and tested unannealed.   The increased tensile strength may not be entirely due to the presence of carbon, for these tests were made on "commercial cobalt".

VIII. The effect of carbon and other impurities in the "commercial cobalt" is, to greatly increase the percentage reduction and elongation, which rises, in most cases, well above 20 per cent.

COMPRESSIVE STRENGTH MEASUREMENTS.

The measurements of the compressive strength of cobalt were made in the same Riehlé universal standard vertical screw power testing machine, of 100,000 lbs. capacity, that was used for the tensile strength measurements, and which is described on page 17, and shown in Plate II.

Test Bars.

All bars for compressive strength measurements were $1\frac{1}{2}$ inches long and $\frac{3}{4}$ inches diameter.

Method of Measurement.

The measurement was made using the testing machine in the standard manner, with a chart in place as shown in Plate II, to get an autographic stress-strain diagram for each sample.

The form of these stress-strain diagrams, obtained on all of our compression samples, is shown in Fig. 6.

From point **O** to point **P**, the compression or strain of the material was very slight, and was proportional to the stress or load applied. The yield point is not very clearly differentiated from the proportional limit on these diagrams, and has been taken to be at point **P**. **M** is the maximum stress, and **R** the rupture stress, or compression breaking load, which values are given in our tables. Below are tables of the data for these measurements, as obtained by us on "commercial cobalt", and on pure cobalt.

Compressive Strength Measurements of "Commercial Cobalt."

Cast and Unannealed.

Sample number.	Date.	Heat and mechanical treatment.	Analysis.	Compressive breaking load in lbs. per square inch.	Yield point in pounds per square inch.	Remarks.
H 87 b	Feb. 10, 1913	Cast from just above melting point, allowed to cool in iron mould, and turned in lathe.	Co 98·5 Fe 0·80 Ni 0·20 C 0·37 S 0·014 Si 0·020 Ca none P none	172000	29000	
H 87 c	Jan. 16, 1914	Cast from just above melting point, allowed to cool in iron mould, and turned in lathe.	Co 97·8 Fe 1·46 Ni 0·50 C 0·18 S 0·020 Si 0·020 Ca none P 0·012	178000	52100	Metal medium hard and tough. Machines with medium long curling chip.
H 87 a and e	Dec. 22, 1913	Cast from just above melting point, allowed to cool in iron mould, and turned in lathe.	Co 98·7 Fe 0·80 Ni 0·20 C 0·23 S 0·030 Si 0·020 Ca none P trace	184000	47600	

Cast and Unannealed—(continued).

Sample number.	Date.	Heat and mechanical treatment.	Analysis.	Compressive breaking load in lbs. per square inch.	Yield point in lbs. per square inch.	Remarks.
H 87 d	Feb. 10, 1913	Cast from just above melting point, allowed to cool in iron mould, and turned in lathe.	Co 98·5 Fe 0·80 Ni 0·20 C 0·37 S 0·016 Si 0·020 Ca none P none	184000	31200	
H 109	Dec. 22, 1913	Cast from just above melting point, allowed to cool in iron mould, and turned in lathe.	Co 96·8 Ni 0·56 Fe 2·36 S 0·022 C 0·062 P 0·017	197500	35000	
H 211	Dec. 22, 1913	Cast from just above melting point, allowed to cool in iron mould, and turned in lathe.	S 0·080 C 0·17 P 0·031	92000	41000	Low value due to segregation of impurities. See microphotograph, Plate XIII.
H 130	Dec. 22, 1913	Cast from just above melting point, allowed to cool in iron mould, and turned in lathe.	Co 96·52 Fe 1·27 Ni 2·00 S 0·053 C 0·305 P 0·015	94000	36000	Metal short grained and brittle. Machines with a short chip.

Cast and Annealed.

Sample number.	Date.	Heat and mechanical treatment.	Analysis.	Compressive breaking load in lbs. per square inch.	Yield point in lbs. per square inch.	Remarks.
H 211	Jan. 10, 1913	Cast from just above melting point, allowed to cool in iron mould, and turned in lathe. Annealed from 700° C.	S 0·080 C 0·17 P 0·031	98200	54300	Low value due to segregation of impurities. See microphotograph, Plate XIII.
H 214	Apl. 21, 1914	Cast from just above melting point, allowed to cool in iron mould, and turned in lathe. Annealed at 850° C.	Co 98·71 Fe 1·45 Ni none S 0·012 Mn 2·04 Ca none Si 0·011 P 0·010 C 0·067	144000	40700	Fine grained and uniform.
H 87c	May, 19 1914	Cast from just above melting point, allowed to cool in iron mould, and turned in lathe. Annealed at 850°C.	Co 97·8 Fe 1·46 Ni 0·50 C 0·18 S 0·020 Si 0·020 Ca none P 0·012	124000	56500	
H 87c	June 17, 1914	Cast from just above melting point, allowed to cool in iron mould, and turned in lathe. Annealed at 950°C.	Co 97·8 Fe 1·46 Ni 0·50 C 0·18 S 0·020 Si 0·020 Ca none P 0·012	148000	61000	

Compressive Strength Measurements of Pure Cobalt.

Cast and Unannealed.

Sample number.	Date.	Heat and mechanical treatment.	Analysis.	Compressive breaking load in lbs. per square inch.	Yield point in lbs. per square inch.	Remarks.
H 212	Jan. 10, 1913	Cast from just above melting point, allowed to cool in iron mould, and turned in lathe.	Co 99·9 Ni none Fe 0·20 S 0·017 C none Si none Ca none	88900	40700	Metal soft and brittle. Turned with a very short chip.
H 212	Jan. 24, 1914	Cast from just above melting point, allowed to cool in iron mould, and turned in lathe.	Co 99·9 Ni none Fe 0·20 S 0·017 C none Si none Ca none	104800	40700	
H 213	Jan. 24, 1914	Cast from just above melting point, allowed to cool in iron mould, and turned in lathe.	Co 99·73 Ni none Fe 0·14 S 0·019 C none	154000	36200	
H 193a	Dec. 9, 1913	Cast from just above melting point, allowed to cool in iron mould, and turned in lathe.	Co 99·6 Fe 0·41 S 0·025 C none Ni trace P trace Ca none Si none	135800	39200	
H 193	Jan. 10, 1913	Cast from just above melting point, allowed to cool in iron mould, and turned in lathe.	Co 99·6 Fe 0·41 S 0·025 C none Ni trace P trace Ca none Si none	123900	54200	

Cast and Annealed.

Sample number.	Date.	Heat and mechanical treatment.	Analysis.	Compressive breaking load in lbs. per square inch.	Yield point in lbs. per square inch.	Remarks.
H 212	Jan. 16, 1914	Cast from just above melting point, allowed to cool in iron mould, and turned in lathe. Annealed from 700° C.	Co 99·9 Ni none Fe 0·20 S 0·017 C none Si none Ca none	129100	63400	
H 213	Jan. 16, 1914	Cast from just above melting point, allowed to cool in iron mould, and turned in lathe. Annealed from 700° C.	Co 99·73 Ni none Fe 0·14 S 0·019 C none	114300	65600	Metal soft and brittle. Machines with a short chip.
H 217	May 19, 1914	Cast from just above melting point, allowed to cool in iron mould, and turned in lathe. Annealed at 850°C.	Co 99·20 Ni none Fe 0·730 S 0·016 Ca none Si 0·091 P 0·0077 C 0·036 Al 0·021	102000	45200	

Cast and Annealed—(continued).

Sample number.	Date.	Heat and mechanical treatment.	Analysis.	Compressive breaking load in lbs. per square inch.	Yield point in lbs. per square inch.	Remarks.
H 217	May 20, 1914	Cast from just above melting point, allowed to cool in iron mould, and turned in lathe. Annealed at 850°C.	Co 99·20 Ni none Fe 0·730 S 0·016 Ca none Si 0·091 P 0·0077 C 0·036 Al 0·021	106000	65600	
H 217	June 17, 1914	Cast from just above melting point, allowed to cool in iron mould, and turned in lathe. Annealed at 950°C.	Co 99·20 Ni none Fe 0·730 S 0·016 Ca none Si 0·091 P 0·0077 C 0·036 Al 0·021	134800	40700	

There is, practically, no literature on the compressive strength of cobalt.

We measured the compressive yield point of pure nickel under conditions identical with the above measurements for cobalt, and found it to be 20,000 pounds per square inch. This was for a sample cast from just above the melting temperature, allowed to cool in iron mould, and tested unannealed. The nickel analysed as follows:—

	%
Ni	99·29
Fe	0·48
Co	none
S	0·025
Si	0·042
Ca	none
C	none

Conclusions.

Pure Cobalt.

I. The compressive strength of pure cobalt cast and unannealed is in the neighbourhood of 122,000 pounds per square inch. This is the average of five measurements on cobalt cast from just above its melting point, allowed to cool in iron mould, and machined in lathe to test bars.

II. The effect of annealing on the compressive strength of cast pure cobalt is not very marked—the average of five measurements of the compressive strength of annealed cast cobalt is 117,200 pounds per square inch. There seems to be a tendency to lower the compressive strength by annealing.

III. The compressive yield point of pure cobalt is 56,100 pounds per square inch, for the annealed samples, compared with 42,200 pounds per square inch for the unannealed samples. Thus the yield point seems to be slightly raised by annealing.

IV. The compressive yield point of pure cobalt, cast from just above the melting temperature, allowed to cool in iron mould and machined in lathe to test bar, is considerably greater than either that of pure iron or nickel prepared and tested under the same conditions.

"Commercial Cobalt."

V. The effect of the addition of carbon is to increase the compressive breaking strength of cobalt, the value rising well above 175,000 pounds per square inch by the addition of from 0·060 to 0·30 per cent carbon. These values refer to cobalt cast from just above the melting point, allowed to cool in iron mould, machined in lathe, and tested unannealed. The increased compressive strength may not be entirely due to the presence of carbon, for these tests were made on "commercial cobalt".

VI. The effect of carbon and other impurities in the "commercial cobalt" does not seem greatly to affect the yield point through the range of our observation, although on the average from 0·060 to 0·30 per cent of carbon with the other impurities shown, tends to lower it from 5 to 10 per cent, both annealed and unannealed.

VII. The effect of annealing "commercial cobalt" is to lower its compressive strength, our values averaging 140,000 pounds per square inch for the annealed samples, as compared with 183,000 pounds per square inch for the unannealed samples.

VIII. The compressive yield point of "commercial cobalt" is, similar to that for the pure metal, slightly raised by annealing; our values averaging 39,000 pounds per square inch for unannealed samples, as compared with 53,000 pounds per square inch for the annealed samples.

MACHINING, ROLLING, AND SWAGING OF METALLIC COBALT.

Turning Properties.

Pure metallic cobalt may be readily machined in the lathe, although it is somewhat brittle, and yields a short chip. The addition of small amounts of carbon renders cobalt less brittle, and yields a longer curling chip upon turning.

Swaging of Cobalt.

Cast cobalt of extreme purity, which has been cast either in iron or in sand moulds, whether cooled slowly or rapidly, cannot be directly swaged down to a fine wire without special mechanical-heat treatment.

On the other hand, "commercial cobalt" containing small percentages of carbon as described in this paper, may readily be swaged down from cast bars to wires of any desired diameter.

Swaging Machines.

For our experiments on the swaging of cobalt, we used a No. 3 Dayton swaging machine, manufactured by the Excelsior Needle Company of Torringtown, Conn. This is shown in Plate III.

With this machine the metal is not drawn out, as is the case with wire-drawing machines, but is rather hammered down by being placed within a pair of dies, carried in a slot in the face of a revolving mandrel, and outside of which is an annular rack containing a number of hardened steel rollers. The dies thus revolve rapidly around the work, which is hammered by them as they pass between opposite pairs of rolls on either side of it.

With this machine it is comparatively easy to swage hard steel into fine wires. The steel will pass through the dies either hot or cold. When, however, a bar of pure cobalt, which has been turned in a lathe to give it a smooth uniform surface, was placed in the swaging machine cold, it cracked along its entire length, and broke off at many places. This was repeated several times with different bars of the metal, each time with the same result, showing that the metal could not be swaged cold.

It became evident that pure cobalt must be given some heat treatment before it could be swaged at all. Hence a bar was first annealed from a temperature of 700°C, by heating it slowly to this temperature in a gas muffle furnace, holding it there for a short time, and then allowing it to cool with the furnace, during several hours. It was then heated to different temperatures before being placed in the swaging machine, with the following results.

At 900°C the metal crumbled in the machine as though it were extremely hot short, although the sulphur content was as low as 0·018 and 0·020 per cent.

At 700°C to 800°C it still cracked, and broke very badly in the machine.

At 500°C to 600°C, however, the bar could be passed through one or two dies without any apparent cracking. It would not go farther than this, although the bar was reannealed after each pass, and reductions in diameter of only 0·002 to 0·003 inches, on a 3/8″ bar, were made at each pass. At lower temperatures than this the metal would crack still more, and hence it is obvious that the metal must be given some special treatment to render it more ductile before it can be swaged down to a wire.

Cobalt, like iron and certain other metals, will absorb considerable quantities of gases when it is in the molten state; and as the gases in the metal will in all probability have a bad effect on its swaging properties an attempt was made to remove any of these gases that may have been dissolved in the metal and which still remained in the solid bar. With this in view, a bar which had been cast in an iron mould was heated slowly in a good vacuum to a temperature of about 700°C, where it was held for several hours, at the end of which time the bar was allowed to cool slowly in the furnace. This treatment is claimed to have rendered tungsten more ductile; but on attempting to swage a cobalt bar which had been treated thus, very little, if any, improvement was observed in its swaging properties.

The method which finally succeeded, consisted in slowly cooling the bar from a high temperature, 1100°C to 1150°C, under a high pressure. This was accomplished in the following way. The bar was placed within an iron mould, squeezed tightly by means of clamps, and the whole heated slowly to the above temperature. The mould, with bar, was then removed from the furnace, and the outer portion of the mould chilled while the inner portion still remained hot. The consequent contraction through cooling of the outer portion exerted a considerable pressure on the inner hot bar of metal. The cooling under this pressure continued for three or four hours, after which the metal swaged at a dull red heat with very little difficulty.

The process of swaging consisted in passing the metal, heated to a dull red heat, through successive dies, which hammered it down until a wire of the required diameter was obtained. However, the temperature at which the bar was passed through the dies had to be carefully regulated, as the metal apparently would not swage at all when cold, and when hot only between 500° and 600°C. By thus controlling the temperature, and feeding in the bar very slowly, good, smooth, uniform wires were obtained.

Conclusions.

(1). Pure cobalt may be machined in a lathe as readily as pure nickel or pure iron, although it is somewhat brittle and yields a short chip.

(2). "Commercial cobalt", containing small percentages of carbon, machines very readily after the manner of mild steel.

PLATE III.

Swaging Machine.

(3). Cast cobalt of extreme purity cannot be rolled or swaged without developing cracks, unless given a special mechanical heat treatment.

(4). Cast cobalt of extreme purity may be rolled or swaged to any extent by cooling the casting under extreme pressures, followed by passing through rolls or dies at temperatures between 500 to 600°C, and so as to reduce the bar by small percentages at each pass.

(5). "Commercial cobalt" containing small percentages of carbon, may be rolled or swaged from cast bars to any extent, provided that the metal be worked at a red heat.

MEASUREMENT OF ELECTRICAL RESISTANCE.

The potentiometer method of electrical measurement, which is in reality a measurement of the drop in potential along a known length of wire when a definite current is flowing through it, was employed for our measurements on pure and "commercial" cobalt.

The samples of metal which were used were all cast from just above their melting points, allowed to cool slowly in an iron mould, and thereafter swaged down to wires of given diameters according to the method described under "Swaging of Pure Cobalt", page 29.

Analyses of the samples will be given in each instance with the tabulated results of the measurements.

Description of Apparatus.

Fig. 7 is a diagrammatic sketch of the electrical circuits as they were used in the potentiometer method of measuring the electrical resistance of cobalt.

W is a storage battery, two volts, which sends a current through the circuit **WRDAMCBW**, which flows in the direction from **R** to **B**. This circuit is known as the potentiometer circuit. **AC** is a series of fifteen 5 ohm resistance coils, and **CB** is a 5 ohm slide-wire, consisting of several turns of constantine wire mounted on a marble cylinder. **S**td is a cadmium standard cell, electromotive force 1·0189 volts, which bears the certificate of the United States Bureau of Standards. The standard cell is connected from a point **X** in the coils **AC** to the switch **T**, which is set at a point in the resistance **DTA**, such that the electromotive force between **T** and **X**, due to the battery **W**, is exactly equal to that of the standard cell.

This balanced condition is determined by the galvanometer **G**, which is connected in the circuit of the standard cell by throwing the switch **U** into the dotted position. The resistance **R** is adjusted until there is no deflection of the galvanometer, which signifies the balanced condition above mentioned.

The resistance coils from **A** to **C** are a set of fifteen 5 ohm coils, and the point **X** is such that ten of them are included between **A** and **X**. **ATD** is a standard resistance such that there is included between **A** and **T** exactly 0·945 ohms. When the balance was made there flowed, therefore, through the potentiometer circuit, a current $I = \dfrac{1 \cdot 0189}{50 \cdot 945} = \dfrac{1}{50}$ amperes.

This adjustment is made so that, for this current in the potentiometer circuit, the drop in potential across two adjacent coils along **AC** is exactly 1/10 of a volt.

The switch used is now thrown to connect an unknown electromotive force (**E.M.F.** in the diagram) through the galvanometer, and in such a way that the current from the new source flows in the same direction as that from the standard cell. The sliding contact **M**¹ is brought to the zero end of the slide wire, and the moving contact **M** is shifted from **C** towards **A**, step by step, until the galvanometer deflection is reversed in direction. **M** is left at the last point for which a galvanometer deflection is in the same direction as when **M** was at **C**. Then the contact **M**¹ is moved along the slide wire until the galvanometer deflection is zero.

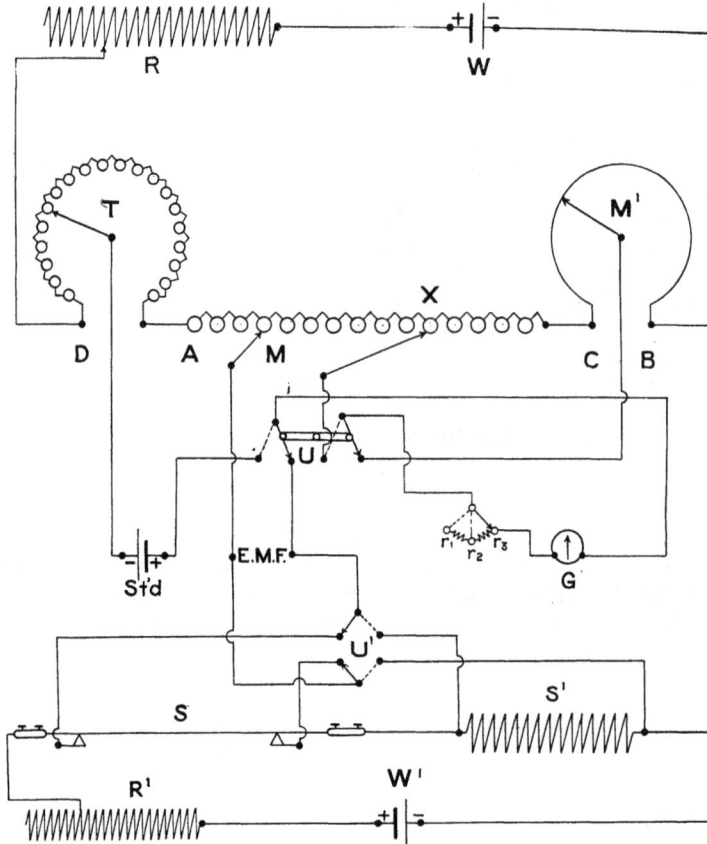

Fig. 7. **Arrangement of circuits for electrical resistance measurements.**

W and **W**¹	= 2 volt and 4 volt storage batteries respectively.
R and **R**¹	= Adjustable rheostats.
U and **U**¹	= Double-throw switches.
G	= Galvanometer.
R and **R**¹	= Small resistances for protection of galvanometer in making adjustments.
Std.	= Standard cell, e.m.f. = 1·0189 volts.
S	= Wire tested.
S¹	= Standard resistance of 0·1 ohms correct to $\frac{1}{25}$ of one per cent, and with no temperature coefficient.

The reading of the contact point **M** gives the value of the electromotive force in tenths of a volt, and that of **M**¹ from hundredths to hundred-thousandths of a volt. Thus a very accurate measurement of the unknown electromotive force is obtained in terms of the known standard.

The unknown electromotive force in these experiments is not a cell, but is the drop in potential along **S**. **S** is a given length of the cobalt wire, whose resistance is to be measured, through which a small current

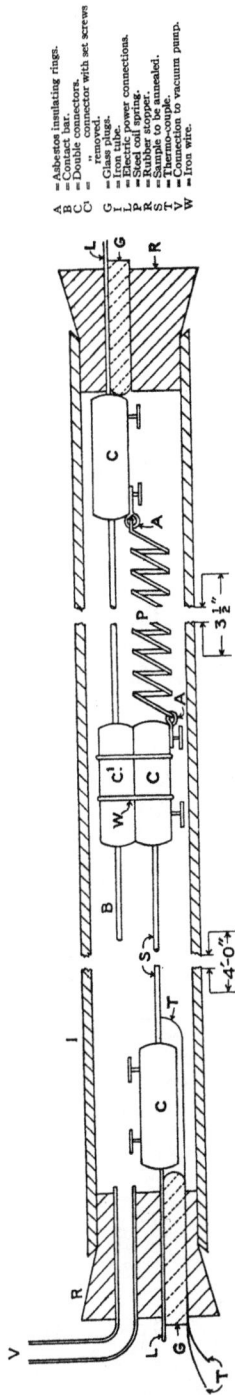

Fig. 8. Annealing furnace (full size): longitudinal section, taken vertically, through centre.

A = Asbestos insulating rings.
B = Contact bar.
C = Double connectors.
C¹ = " connector with set screws removed.
G = Glass plugs.
I = Iron tube.
L = Electric power connections.
P = Steel coil spring.
R = Rubber stopper.
S = Sample to be annealed.
T = Thermo-couple.
V = Connection to vacuum pump.
W = Iron wire.

is passing from battery W^1 connected as shown in the diagram. In this latter circuit W^1S there is also connected a standard resistance S^1 of $0 \cdot 1$ ohms. By throwing the switch U^1 into the dotted position the drop in potential along S^1 was measured. Knowing the drop in potential along S and also along S^1, when the same current is passing through each, the resistances are known from the following equation—

$$\frac{\text{unknown resistance } S}{\text{known resistance } S^1} = \frac{\text{drop in potential along } S}{\text{drop in potential along } S^1}$$

Method of Computation.

The length S between two knife edges, which formed the contact points between which the electromotive force was determined, was carefully measured to tenths of a millimeter. The average diameter of the wire was measured to thousandths of a millimeter, and from these data the specific resistance of the wire in ohms per cubic centimeter was calculated to be—

$$R = \mu \frac{1}{A} \qquad \text{or} \qquad \mu = \frac{R A}{1}$$

where
- R = total resistance of S in ohms.
- 1 = length of S in centimeters.
- A = average cross section of S in square centimeters.
- μ = specific resistance in ohms per centimeter cube.

After this measurement the wire was cut to the exact length S, carefully weighed to the nearest milligram, and the density of the wire determined by the Archimedes principle. From these data the resistance of the wire in ohms per meter gram was calculated as follows:—

$$R = \mu \frac{1}{A}$$

$$D = \frac{M}{V} = \frac{M,}{1A} \quad \text{or} \quad A = \frac{M}{Dl},$$

from which we have—

$$R = D\mu \frac{1^2}{M} = \frac{kl^2}{M}, \quad \text{or} \quad k = \frac{Rm}{1^2}, \text{ where,}$$

- R = resistance of S in ohms,
- M = mass of S in grams,
- D = density of S,
- k = specific resistance of S in ohms per meter gram,
- V = volume of S in cubic centimeters.

For a comparison of k and μ it should be noted that k is equal to μ multiplied by the density of the wire times 10^4.

Below is a table of the values obtained by this type of measurement both in ohms per centimeter cube, and in ohms per meter gram, together with analyses of the specimens. In each of the values for k, the computations were made directly from the length of the wire in meters and the mass of the wire in grams.

Annealing of Wires.

The effect of annealing on the conductivity of both pure and "commercial" cobalt was studied. In this connexion the annealing was accomplished by two methods:

(1). Passing a suitable electric current through the wire in vacuo.

(2). Heating within an electric furnace in a CO_2 atmosphere.

The annealing furnace used for heating in vacuo consisted of a cylindrical glass tube about 4 feet in length and 2 inches in diameter, and sealed off at the end with rubber stoppers. Through the ends protruded copper leads and a connexion to a vacuum pump. The slack in the cobalt wire, developed upon heating, was taken up by a coiled spring, as shown in Fig. 8. The approximate temperature was measured by a thermocouple placed against the annealing wire.

The furnace used for annealing in a carbon dioxide atmosphere consisted of an iron tube about 4 feet long, and 2 inches diameter, wound with suitable insulated nichrome wire. The ends, which were sealed off with rubber stoppers, were water cooled, and contained a suitable gas inlet and outlet.

In the tables following, showing the results of these resistance measurements, in the column under "Remarks", will be indicated by which of the above methods, "Vacuum Furnace" or "Carbon Dioxide Atmosphere", the annealing was accomplished.

Electrical Resistance of "Commercial Cobalt."

All of the samples of the following table of measurements were cast from just above the melting point, allowed to cool in an iron mould, and turned in a lathe to a bar of about one centimeter in diameter. They were then passed through the swaging machine, after special heat and mechanical treatment, as described on page 29, and drawn down to wires of the desired diameter. They were not annealed or given further heat treatment after drawing.

Unannealed.

Sample number.	Date.	Temperature of wire in degrees centigrade.	Length of wire in cms.	Average cross sectional area of wire in sq. cms.	Weight of wire in grams.	Analysis.	Resistance in ohms per centimeter cube.	Resistance in ohms per meter gram.
H 192	Nov. 7 1913.	21.2	83.25	0.005568	3.9881	Co 99.63 Ni none Fe 0.60 S 0·023 C 0.090 Si trace Ca trace	229.6x10^{-7}	1.977
H 192	Nov. 7 1913.	21.0	50·20	0·005568	2·4051	Co 99.63 Ni none Fe 0·60 S 0·023 C 0.090 Si trace Ca trace	231.2x10^{-7}	1·992
H 192	Nov. 7 1913.	21·8	50·20	0·005568	2·4051	Co 99·63 Ni none Fe 0·60 S 0·023 C 0.090 Si trace Ca trace	231·5x10^{-7}	1·993
H 193	Nov. 8 1913.	21·9	75·07	0·01885	12·3871	Co 92·36 Ni 2·73 Fe 4·49 S 0·018 C none	144·4x10^{-7}	1·271
H 193	Nov. 8 1913.	22·0	51·67	0·01885	8·526	Co 92·36 Ni 2·73 Fe 4·49 S 0·018 C none	144·5x10^{-7}	1·271

Unannealed—(continued).

Sample Number.	Date.	Temperature of wire in degrees centigrade.	Length of wire in cms.	Average cross sectional area of wire in sq. cms.	Weight of wire in grams.	Analysis.		Resistance in ohms per centimeter cube.	Resistance in ohms per meter gram.
H 214	Apl. 3 1914.	22·5	74·95	0·006374	4·304	Co Ni Fe Si Ca S C P	98·71 none 1·15 0·14 none 0·012 0·039 0·010	$105·8 \times 10^{-7}$	0·9530
H 214	Apl. 3 1914.	22·5	74·95	0·006374	4·304	Co Ni Fe Si Ca S C P	98·71 none 1·15 0·14 none 0·012 0·039 0·010	$105·8 \times 10^{-7}$	0·9530
H 214	Apl. 3 1914.	22·0	52·64	0·006419	3·023	Co Ni Fe Si Ca S C P	98·71 none 1·15 0·14 none 0·012 0·039 0·010	$105·1 \times 10^{-7}$	0·9545
H 214	Apl. 3 1914.	22·0	52·64	0·006319	3·023	Co Ni Fe Si Ca S C P	98·71 none 1·15 0·14 none 0·012 0·039 0·010	$104·8 \times 10^{-7}$	0·9524
H 214	Apl. 6 1914.	16·0	82·56	0·006319	4·723	Co Ni Fe Si Ca S C P	98·71 none 1·15 0·14 none 0·012 0·039 0·010	$104·2 \times 10^{-7}$	0·9431
H 214	Apl. 6 1914.	15·5	82·56	0·006319	4·723	Co Ni Fe Si Ca S C P	98·71 none 1·15 0·14 none 0·012 0·039 0·010	$103·8 \times 10^{-7}$	0·9403
H 214	Apl. 7 1914.	18·0	59·95	0·006305	3·429	Co Ni Fe Si Ca S C P	98·71 none 1·15 0·14 none 0·012 0·039 0·010	$104·7 \times 10^{-7}$	0·9502
H 214	Apl. 7 1914.	18·0	59·95	0·006305	3·429	Co Ni Fe Si Ca S C P	98·71 none 1·15 0·14 none 0·012 0·039 0·010	$104·7 \times 10^{-7}$	0·9502

Annealed.

All the samples of the following table were annealed at the temperatures given in the last column.

Sample number.	Date.	Temperature of wire in degrees centigrade.	Length of wire in ohms	Average cross sectional area of wire in sq. cms.	Weight of wire in grams.	Analysis.	Resistance in ohms per centimeter cube.	Resistance in ohms per meter gram.	Remarks.
H 214	Apl. 11 1914	13·0	74·65	0·006333	4·260	Co 98·71 Ni none Fe 1·15 Si 0·14 Ca none S 0·012 C 0·039 P 0·010	$91·76 \times 10^{-7}$	0·8286	Annealed at 350°C for about 5 hours by passing current through wire. Vacuum furnace.
H 214	Apl. 11 1914.	13·0	47·50	0·006376	2·713	Co 98·71 Ni none Fe 1·15 Si 0·14 Ca none S 0·012 C 0·039 P 0·010	$91·49 \times 10^{-7}$	0·8205	Annealed at 350°C for approximately 5 hours, by passing current thro' wire. Vacuum furnace.
H 214*	Apl. 21 1914.	21·5	83·69	0·006461	4·845	Co 98·71 Ni none Fe 1·15 Si 0·14 Ca none S 0·012 C 0·039 P 0·010	$104·1 \times 10^{-7}$	0·9325	Annealed at 200°C for approximately 5 hours in furnace, CO_2 atmosphere.
H 214*	Apl. 21, 1914.	21·5	49·97	0·006461	2·888	Co 98·71 Ni none Fe 1·15 Si 0·14 Ca none S 0·012 C 0·039 P 0·010	$103·7 \times 10^{-7}$	0·9264	Annealed at 200°C for approximately 5 hours in furnace, CO_2 atmosphere.
H 214*	Apl. 22, 1914.	19·2	82·80	0·006447	4·799	Co 98·71 Ni none Fe 0·15 Si 0·14 Ca none S 0·012 C 0·039 P 0·010	$102·6 \times 10^{-7}$	0·9225	Annealed at 300°C for 2 hours in furnace, CO_2 atmosphere.
H 214*	Apl. 22, 1914	19·2	45·86	0·006447	2·657	Co 98·71 Ni none Fe 0·15 Si 0·14 Ca none S 0·012 C 0·039 P 0·010	$102·9 \times 10^{-7}$	0·9249	Annealed at 300°C for 2 hours in furnace, CO_2 atmosphere.
H 214*	Apl. 23, 1914.	19·0	82·40	0·006390		Co 98·71 Ni none Fe 1·15 Si 0·14 Ca none S 0·012 C 0·039 P 0·010	$100·7 \times 10^{-7}$	0·9090	Annealed at 400°C for 2 hours in furnace, CO_2 atmosphere.
H 214*	Apl. 23, 1914.	19·0	54·94	0·006404		Co 98·71 Ni none Fe 1·15 Si 0·14 Ca none S 0·012 C 0·039 P 0·010	$100·2 \times 10^{-7}$	0·9019	Annealed at 400°C for 2 hours in furnace, CO_2 atmosphere.

*Samples H 214 series, Apl. 21-23, are all the same wire, annealed and unannealed.

Annealed—(continued).

Sample number.	Date.	Temperature of wire in degrees centigrade.	Length of wire in ohms	Average cross sectional area of wire in sq. cms.	Weight of wire in grams.	Analysis	Resistance in ohms per centimeter cube.	Resistance in ohms per meter grams.	Remarks.
H 214	Apl. 24, 1914.	18·5	78·11	0·006291	4·478	Co 98·71 Ni none Fe 1·15 Si 0·14 Ca none S 0·012 C 0·039 P 0·010	98.42×10^{-7}	0·8969	Annealed at 500°C for 2 hours in furnace, CO_2 atmosphere.
H 214	Apl. 24, 1914.	18·5	56·35	0·006291	3·230	Co 98·71 Ni none Fe 1·15 Si 0·14 Ca none S 0·012 C 0·039 P 0·010	98.14×10^{-7}	0·8941	Annealed at 500°C for 2 hours in furnace, CO_2 atmosphere.
H 214(a)	Apl. 25, 1914.	19·3	81·65	0·006319	4·664	Co 98·71 Ni none Fe 1·15 Si 0·14 Ca none S 0·012 C 0·039 P 0·010	93.34×10^{-7}	0·8437	Annealed at 600°C for 1 hour in furnace, CO_2 atmosphere.
H 214	Apl. 25, 1914.	19·3	57·50	0·006319	3·285	Co 98·71 Ni none Fe 1·15 Si 0·14 Ca none S 0·012 C 0·039 P 0·010	93.63×10^{-7}	0·8466	Annealed at 600°C for 1 hour in furnace, CO_2 atmosphere.
H 214	Apl. 27, 1914.	15·6	66·67	0·006291	3·801	Co 98·71 Ni none Fe 1·15 Si 0·14 Ca none S 0·012 C 0·039 P 0·010	91.44×10^{-7}	0·8285	Annealed at 700°C for 1 hour in furnace, CO_2 atmosphere.
H 214	Apl. 27, 1914.	15·8	51·01	0·006291	2·907	Co 98·71 Ni none Fe 1·15 Si 0·14 Ca none S 0·012 C 0·039 P 0·010	90.90×10^{-7}	0·8232	Annealed at 700°C for 1 hour in furnace, CO_2 atmosphere.
H 214	Apl. 28, 1914.	18·4	53·55	0·006263	3·046	Co 98·71 Ni none Fe 1·15 Si 0·14 Ca none S 0·012 C 0·039 P 0·010	90.35×10^{-7}	0·8204	Annealed at 800°C for 1 hour in furnace, CO_2 atmosphere.

(a) Note drop in resistance between 500°C and 600°C.

Electrical Resistance of Pure Cobalt

All of the samples of the following table of measurements were cast from just above the melting point, allowed to cool in an iron mould, and turned in a lathe to a bar of about one centimeter in diameter. They were then passed through the swaging machine, after special heat and mechanical treatment, as described on page 29, and drawn down to wires of the desired diameter. They were not annealed or given further heat treatment after drawing.

Unannealed.

Sample number.	Date.	Temperature in degrees centigrade.	Length of wire.	Average cross sectional area in sq. cms.	Weight in grams.	Analysis.	Resistance in ohms per centimeter cube.	Resistance in ohms per meter gram.
H 212	Jan. 22, 1914.	17°C.	92·81	0·005890	4·863	Co 99·9 Ni none Fe 0·20 S 0·017 C none Si none Ca none	$87 \cdot 27 \times 10^{-7}$	0·7766
H 212	Jan. 22, 1914.	17	92·81	0·005890	4·863	Co 99·9 Ni none Fe 0·20 S 0·017 C none Si none Ca none	$88 \cdot 04 \times 10^{-7}$	0·7834
H 212	Jan. 22, 1914.	17	50·23	0·005890	2·631	Co 99·9 Ni none Fe 0·20 S 0·017 C none Si none Ca none	$88 \cdot 08 \times 10^{-7}$	0·7837
H 212	Jan. 22, 1914.	17	50·23	0·005890	2·631	Co 99·9 Ni none Fe 0·20 S 0·017 C none Si none Ca none	$88 \cdot 22 \times 10^{-7}$	0·7840
H 212b	Jan. 22, 1914.	14	79·41	0·005822	4·113	Co 99·9 Ni none Fe 0·20 S 0·017 C none Si none Ca none	$86 \cdot 66 \times 10^{-7}$	0·7713
H 212b	Jan. 22, 1914.	14	79·41	0·005822	4·113	Co 99·9 Ni none Fe 0·20 S 0·017 C none Si none Ca none	$86 \cdot 80 \times 10^{-7}$	0·7727
H 212b	Jan. 22, 1914.	14	48·98	0·005822	2·538	Co 99·9 Ni none Fe 0·20 S 0·017 C none Si none Ca none	$85 \cdot 80 \times 10^{-7}$	0·7637
H 212b	Jan. 22, 1914.	14	48·98	0·005822	2·538	Co 99·9 Ni none Fe 0·20 S 0·017 C none Si none Ca none	$85 \cdot 55 \times 10^{-7}$	0·7616
H 215	Apl. 2, 1914.	22	95·38	0·006547	5·432	Co 99·6 Ni none Fe 0·19 S 0·012 C none Si 0·084 Ca none P 0·0066	$89 \cdot 17 \times 10^{-7}$	0·7756

Unannealed—(continued).

Sample number.	Date.	Temperature in degrees centigrade.	Length of wire.	Average cross sectional area in sq. cms.	Weight in grams.	Analysis.	Resistance in ohms per centimeter cube.	Resistance in ohms per meter gram.
H 215	Apl. 4, 1914.	22	95·38	0·006547	5·432	Co 99·6 Ni none Fe 0·19 S 0·012 C none Si 0·084 Ca none P 0·0066	$89·98 \times 10^{-7}$	0·7826
H 215	Apl. 4, 1914.	23·0	95·38	0·006547	5·432	Co 99·6 Ni none Fe 0·19 S 0·012 C none Si 0·084 Ca none P 0·0066	$89·98 \times 10^{-7}$	0·7826
H 215	Apl. 4, 1914.	23·0	95·38	0·006547	5·432	Co 99·6 Ni none Fe 0·19 S 0·012 C none Si 0·084 Ca none P 0·0066	$90·26 \times 10^{-7}$	0·7852

Annealed.

The following bars were annealed at the temperatures given in the last column.

Sample number.	Date.	Temperature in degrees centigrade.	Length of wire.	Average cross sectional area in sq. cms.	Weight in grams.	Analysis.	Resistance in ohms per centimeter cube.	Resistance in ohms per meter gram.	Remarks.
H 215	Apl. 13, 1914.	17·7	85·21	0·006475	4·834	Co 99·6 Ni none Fe 0·19 S 0·012 C none Si 0·084 Ca none P 0·0066	$85·03 \times 10^{-7}$	0·7449	Annealed at 350°C for approximately five hours by passing current through the wire in vacuum.
H 215	Apl. 13, 1914.	17·7	85·21	0·006475	4·834	Co 99·6 Ni none Fe 0·19 S 0·012 C none Si 0·084 Ca none P 0·0066	$85·18 \times 10^{-7}$	0·7463	Annealed at 350°C for approximately five hours by passing current through the wire in vacuum.
H 215	Apl. 13 1914.	17·7	50·2	0·006475	2·848	Co 99·6 Ni none Fe 0·19 S 0·012 C none Si 0·084 Ca none P 0·0066	$84·92 \times 10^{-7}$	0·7424	Annealed at 350°C for approximately five hours by passing current through the wire in vacuum.
H 215	Apl. 13 1914.	17·7	50·2	0·006475	2·848	Co 99·6 Ni none Fe 0·19 S 0·012 C none Si 0·084 Ca none P 0·0066	$85·04 \times 10^{-7}$	0·7436	Annealed at 350°C for approximately five hours by passing current through the wire in vacuum.
H 215	Apl. 21, 1914	19·0	74·44	0·006151	4·224	Co 99·6 Ni none Fe 0·19 S 0·012 C none Si 0·084 Ca none P 0·0066	$97·42 \times 10^{-7}$	0·8988	Annealed at 400°C for 2 hours in furnace, CO_2 atmosphere.

Annealed—(continued).

Sample number.	Date.	Temperature of wire in degrees centigrade.	Length of wire in ohms.	Average cross sectional area of wire in sq. cms.	Weight of wire in grams.	Analysis.	Resistance in ohms per centimeter cube.	Resistance in ohms per meter gram.	Remarks.
H 215	Apl. 21 1914	18·8	39·89	0·006221	2·264	Co 99·6 Ni none Fe 0·19 S 0·012 C none Si 0·084 Ca none P 0·0066	$97·94 \times 10^{-7}$	0·8938	Annealed at 400°C for 2 hours in furnace, CO_2 atmosphere.
H 215	Apl. 22 1914	21·2	73·99	0·006249	4·162	Co 99·6 Ni none Fe 0·19 S 0·012 C none Si 0·084 Ca none P 0·0066	$101·0 \times 10^{-7}$	0·9108	Annealed at 500°C for 1½ hours in furnace, CO_2 atmosphere.
H 215	Apl. 22 1914	21·2	49·41	0·006249	2·779	Co 99·6 Ni none Fe 0·19 S 0·012 C none Si 0·084 Ca none P 0·0066	$101·1 \times 10^{-7}$	0·9176	Annealed at 500°C for 1½ hours in furnace, CO_2 atmosphere.
H 215(a)	Apl. 23 1914	15·4	72·75	0·006249	4·080	Co 99·6 Ni none Fe 0·19 S 0·012 C none Si 0·084 Ca none P 0·0066	$93·54 \times 10^{-7}$	0·8395	Annealed at 600°C for 1 hour in furnace, CO_2 atmosphere.
H 215	Apl. 23 1914	15·4	52·23	0·006221	2·930	Co 99·6 Ni none Fe 0·19 S 0·012 C none Si 0·084 Ca none P 0·0066	$93·85 \times 10^{-7}$	0·8460	Annealed at 600°C for 1 hour in furnace, CO_2 atmosphere.
H 215	Apl. 24 1914	19·0	51·37	0·006235	2·875	Co 99·6 Ni none Fe 0·19 S 0·012 C none Si 0·084 Ca none P 0·0066	$94·06 \times 10^{-7}$	0·8442	Annealed at 700°C for 1 hour in furnace, CO_2 atmosphere.
H 215	Apl. 24 1914	19·0	74·05	0·006291	4·145	Co 99·6 Ni none Fe 0·19 S 0·012 C none Si 0·084 Ca none P 0·0066	$93·33 \times 10^{-7}$	0·8360	Annealed at 700°C for 1 hour in furnace, CO_2 atmosphere.
H 215	Apl. 28 1914	16·5	73·11	0·006207	4·080	Co 99·6 Ni none Fe 0·19 S 0·012 C none Si 0·084 Ca none P 0·0066	$91·36 \times 10^{-7}$	0·8203	Annealed at 800°C for ½ hour in furnace, CO_2 atmosphere.
H 215	Apl. 28 1914	16·5	54·15	0·006207	3·022	Co 99·6 Ni none Fe 0·19 S 0·012 C none Si 0·084 Ca none P 0·0066	$91·65 \times 10^{-7}$	0·8240	Annealed at 800°C for ½ hour in furnace, CO_2 atmosphere.

(a) Note drop in resistance between 500°C and 600°C, as before.

The following values of the specific resistance of cobalt are taken from the literature:—

Specific Electrical Resistance of Cobalt.

	Temperature	Specific resistance in ohms per centimeter cube.
Copaux [1]	Room	55×10^{-7}
Reuer and Kaneko [2]	Room	64×10^{-7}
Hofman [3]	Room	97×10^{-7}
Knott, C. G. [4]	100°C	121×10^{-7}
Knott. C. G [4]	200°C	159×10^{-7}
Reichardt [5] 99·8% Co	20°C	97×10^{-7}

In addition to these, the following values of the specific resistance of nickel are taken from the literature:—

Specific Electrical Resistance of Nickel.

	Temperature	Specific resistance in ohms per centimeter cube.
Copaux [1]	Room	64×10^{-7}
Reuer and Kaneko [2]	Room	$77 \cdot 2 \times 10^{-7}$
Hofman. [3]	Room	70×10^{-7}
Fleming [6]	0°C	69×10^{-7}
Niccolai [7]	0°C	119×10^{-7}
Harrison, E. P. [8]	0°C	103×10^{-7}

[1] Annalen de Chimie et de Physique (8), Vol. 6, 1905, p. 508.
[2] Ferrum. Vol. 10, p. 257. Chemical Abstracts, 113, p. 3591.
[3] General Metallurgy, 1913, p. 29.
[4] Proceedings of the Royal Society of Edinburgh, 18, 303, 1891.
[5] Annalen de Physic, (4) 6, 832, 1901.
[6] Proceedings of the Royal Society, Vol. 66. p. 50, 1900.
[7] Lincei Rend, 16, (1), p. 757, 1906.
  Lincei Rend, 16, (2), p. 185, 1907.
[8] Proceedings of the Physical Society, Vol. 18, p. 57, 1902.
  Philosophical Magazine, (6) Vol. 3, p. 177, 1902.

Conclusions.

Pure Cobalt.

(1). The specific electrical resistance of cobalt wires of extreme purity is $89 \cdot 64 \times 10^{-7}$ ohms per centimeter cube, or $0 \cdot 7769$ ohms per meter gram, at 18°C. This is the average of twelve observations agreeing well among themselves, and is for wires unannealed after swaging. This is approximately five times that of pure copper.

(2). The effect of annealing cobalt wire of extreme purity in vacuo, at about 350°C for several hours, by passing an electric current through the wire is to diminish its electrical resistance by about 5 per cent. This is not as much as is true of some metals, as for example, aluminium, which diminishes its resistance about 10 per cent by annealing for 2 hours at 250°C.[1]

[1] H. Gwercke, Electrician, Vol. 72, p. 450.
  Chemical Abstracts, 1914, p. 1049.

(3). The effect of annealing cobalt wires of extreme purity in an atmosphere of carbon dioxide gas by heating from an external source is at first to increase the resistance, but with continued annealing at increasingly higher temperatures up to 800°C, the specific resistance drops again. It is particularly noticeable that there is a drop of about 7 per cent in the specific electrical resistance of cobalt wire of extreme purity, annealed in a carbon dioxide atmosphere at 600°C compared with one similarly annealed at 500°C. This drop was from 101×10^{-7} ohms per centimeter cube to $93 \cdot 5 \times 10^{-7}$ ohms per centimeter cube.

(4). The fact that annealing in vacuo diminishes the electrical resistance of pure cobalt, whereas annealing in an inert gas at low temperatures increases its electrical resistance, which is again lowered by heating at higher temperatures, tends to show that the specific electrical resistance of cobalt is largely influenced by the presence of occluded or absorbed gases.

"Commercial Cobalt."

(5). The specific electrical resistance of cobalt, similar to copper and most other metals, increases tremendously by the addition of small percentages of impurities. Less than $0 \cdot 5$ per cent of impurities may treble the electrical resistance.

(6). The specific electrical resistance of "commercial cobalt" varies between 231×10^{-7} and 103×10^{-7} ohms per centimeter cube, for the cases we have studied, depending upon the nature of the small percentages of impurities present. These figures are for wires unannealed after swaging.

(7). The effect of annealing "commercial cobalt" by passing an electric current through the wire in vacuo, is to greatly reduce its specific electrical resistance. Annealing in this way at 350°C for 5 hours reduced the specific resistance by approximately 14 per cent.

(8). The effect of annealing "commercial cobalt" in an atmosphere of carbon dioxide gas by heating from an external source is in general to decrease its resistance. Similar to the case of pure cobalt, there is a sharp decrease in resistance in the sample annealed in an atmosphere of CO_2 gas at 600°C compared with that similarly annealed at 500°C.

These conclusions all refer to measurements on wires made from bars cast from just above the melting temperature, allowed to cool in an iron mould, and then swaged to wires of approximately $0 \cdot 03''$ diameter, in the manner described on pages 29-30.

Magnetic Permeability.

Cobalt is magnetic at all temperatures up to about 1100°C. The magnetic permeability and hysterisis of pure and "commercial" cobalt have been and are being studied at this laboratory, and will be reported in detail in connexion with the publication concerning certain magnetic alloys of cobalt.

Specific Heat Measurements.

Method and Apparatus.

The specific heat of cobalt was determined by the method of mixtures, and the result is probably accurate to within $0 \cdot 5$ per cent.

The method employed consisted in heating a weighed amount of metallic cobalt in the form of short pieces of wire to 100°C, by bringing them into temperature equilibrium with steam at normal temperature and pressure, at the same time having them enclosed so that they were perfectly dry. This was accomplished by a simple boiler device.

When the metal was thus brought to 100°C—which temperature was read on a suitable thermometer, and after constant temperature readings on this thermometer had been obtained for a period of minutes—it was dropped directly from the heater into a suitable calorimeter. Prior to dropping the cobalt at 100°C into the calorimeter, preliminary temperature readings of the water in the calorimeter were made over a period of minutes. These readings were continued with uniform stirring of the calorimeter liquid, after the introduction of the cobalt, until the final equilibrium calorimeter temperature had been reached.

The thermometer was read to one hundredth of a degree centigrade, and readings were taken every twenty seconds. This method obviously gives us the mean specific heat between 100°C and room temperature, approximately 15°C.

As this is a more or less standard calorimeter observation, it is not seen fit to give the minor details of the apparatus and method of procedure.

Material.

The cobalt used for the specific heat measurements analysed as follows:

H 213. Co...99·73
        Ni... none
        Fe... 0·14
        S.. 0·019
        C.. none
        Si... 0·040
        Ca... none
        P.. none

The mean of a series of specific heat measurements, made as indicated above, gives us as the

mean specific heat of cobalt between 15 - 100°C = 0·1053,
with an average deviation of single observations from the mean of about 0·5 per cent.

The writers wish to acknowledge their indebtedness to Professor W. J. Drisko of the Department of Physics, Massachusetts Institute of Technology, Boston, who was good enough to have several specific heat measurements of the same material made under his direction at the Massachusetts Institute of Technology. These measurements were made on the same material as above (H 213), which was sent from this laboratory for the purpose. Following are the measurements:—

H 213, March 18, 1914.

Specific heat of Cobalt.		Deviation of a single observation from the mean.	
	0·1070		0·0014
	0·1037		0·0019
	0·1058		0·0002
	0·1060		0·0004
Mean	0·1056	Average	0·0010

Mean specific heat of cobalt between 15–100°C = 0·1056 ∓ 0·0005.

The following values of the specific heat of cobalt are taken from the literature.

Name of observer.	Temperature range.	Specific heat.
Tilden [1]	—182 to 15C	0·0822
	—78 " 15	0·0939
	15 " 100	0·1030
	15 " 185	0·1047
	15 " 350	0·1087
	15 " 435	0·1147
	15 " 550	0·1209
	15 " 630	0·1234
Tilden [2]	20 " 200°C	0·104
Copaux [3]	500°C	0·1451
	800	0·1846
	1000	0·204
Copaux [4]	20 " 100°C	0·104
Kalmus and Harper[5]	15 " 100°C	0·1053
Drisko[5]	15 " 100°C	0·1056

[1] Proceedings of Royal Society, Vol. 66 1900, p. 244.
[2] Proceedings of Royal Society, Vol. 71, 1903, p. 220.
[3] Comptes Rendus, Vol. 140, 1905, p. 657.
[4] Annalen de Chimie et de Physique (8), Vol, 6. 1905, p. 508.
[5] The Physical Properties of the Metal Cobalt, 1914. (This Report).

From these figures the true specific heat at any temperature may be computed from 0°C to 890°C.

Specific heat $= 0·1058 + 0·0000457t + 0·000000066t^2$.

MICROPHOTOGRAPHS

Following are a series of microphotographs taken in connexion with the metals discussed in the preceding pages. No attempt has been made to make a minute or complete microphotographic study or analysis of cobalt with the small percentages of impurities with which we have to do in this paper. Such a study would be decidedly interesting, but unfortunately, is not possible in the time at our disposal.

The microphotographs shown are rather characteristic, and require no further explanation than the legend accompanying them and reference to the text to which they belong.

Pure Cobalt H 212

Tensile strength bar.
Cast from just above the melting point, allowed to cool in iron mould, and
  turned in lathe.
Date of microphotograph—May 8, 1914.
Density—8·7562 at 17°C

Analysis— %
Co....................................99·9
Fe.................................... 0·20
Ni.................................... none
C..................................... none
S..................................... 0·017
Ca.................................... none
Si.................................... none

Brinell hardness—128·7.
Tensile breaking load—37,900 pounds per square inch.
Melting point—1,478°C, $\mp$ 1·1°C.
Etching—strong iodine for 3 minutes.
Magnification—130 diameters.
Exposure—1 second.

PLATE IV.

Pure Cobalt (H 212).

PLATE V.

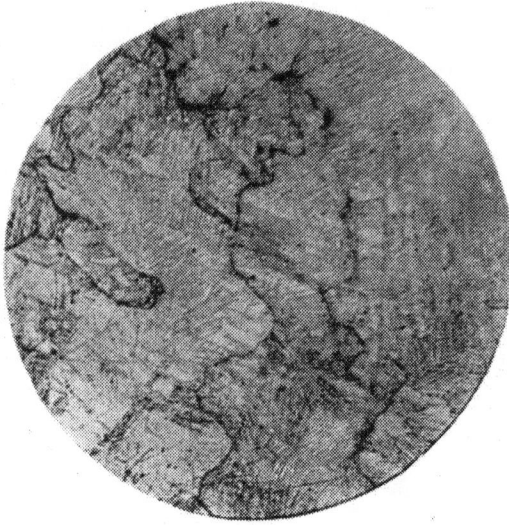

Commercial Cobalt (H 213).

PLATE VI.

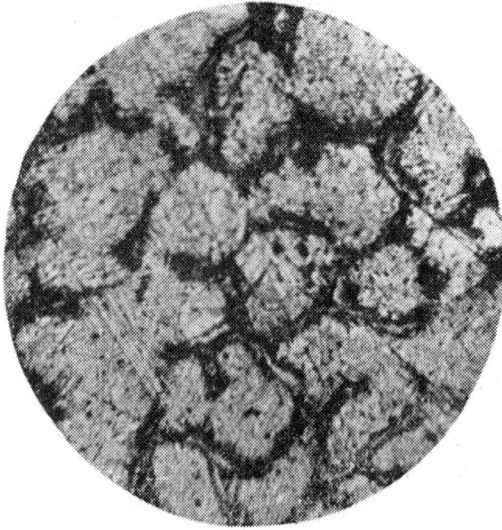

Pure Cobalt (H 214).

PLATE VII.

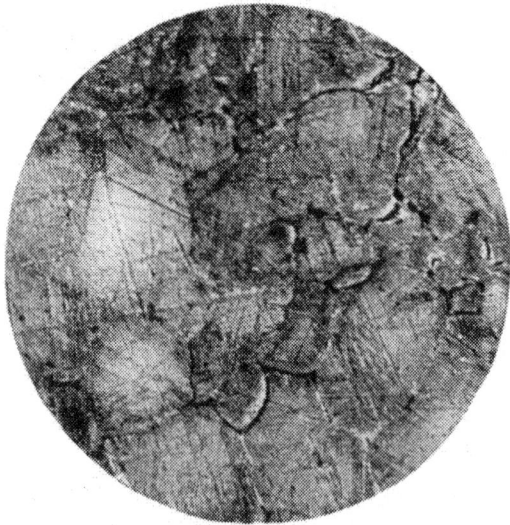

Commercial Cobalt (H 130).

PLATE VIII.

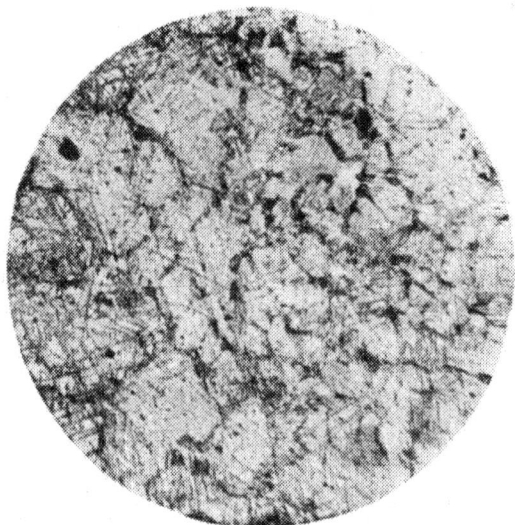

Commercial Cobalt (H 214c).

PLATE IX.

Commercial Cobalt (H 214c).

PLATE X.

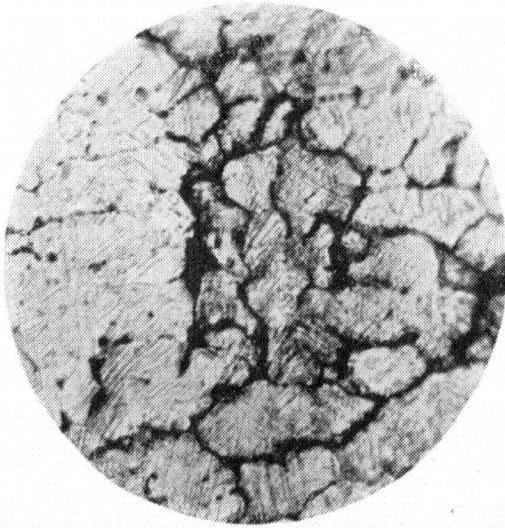

Commercial Cobalt (H 87c).

PLATE XI.

Commercial Cobalt (H 87c).

PLATE XII.

Commercial Cobalt (H 109).

PLATE XIII.

Commercial Cobalt (H 211).

"Commercial Cobalt" H 213

Date of microphotograph—May 9, 1914.
Density—8·7732 at 16°C.

Analysis—

	%
Co	99·73
Ni	none
Fe	0·14
S	0·019
C	none
Si	0·020

Brinell hardness—121·0.
Tensile breaking load—45,300 pounds per square inch.
Etching—strong iodine for 5 seconds.
Magnification—130 diameters.
Exposure—1 second.

Cobalt H 214

Tensile strength bar.
Cast from just above melting point, allowed to cool in iron mould and
    turned in lathe.
Date of microphotograph—May 1, 1914.
Density—8·8490 at 15°C.

Analysis—

	%
Co	98·7
Ni	none
Fe	1·15
Si	0·14
Ca	none
S	0·012
C	none
P	0·011

Etching—strong iodine—14 minutes.
Magnification—130 diameters.
Exposure—2 seconds.
This sample shows polyhedral crystalline structure, with impurities re-
    jected to the boundaries of the crystalline grains.

"Commercial Cobalt" H 130

Tensile strength bar.
Cast from just above the melting point, allowed to cool in iron mould, and
    turned in lathe.
Date of microphotograph—May 1, 1914.
Density—8·7690 at 17°C.

Analysis—

	%
Co	96·5
Ni	2·0
Fe	1·27
C	0·305
S	0·054
P	0·015

Brinell hardness—116·6.
Etching—strong iodine— 7 minutes.
Magnification—130 diameters.
Exposure—2 seconds.

"Commercial Cobalt" H 214c

Tensile strength bar.

Cast from just above the melting point, allowed to cool in iron mould, and turned in lathe. Annealed at 850°C.

Date of microphotograph—May 6, 1914.

Analysis— %

Co.....................................97·09
Ni..................................... none
Fe..................................... 1·45
C..................................... 0·067
S..................................... 0·012
Mn..................................... 2·04
Si..................................... 0·011
Cu..................................... none
P..................................... 0·010

Tensile breaking load—70,500 pounds per square inch.

Etching—strong iodine—5 minutes.

Magnification—130 diameters.

Exposure—1 second.

"Commercial Cobalt" H 214c

Tensile strength bar.

Cast from just above the melting point, allowed to cool in iron mould, and turned in lathe. Annealed at 1,000°C.

Date of microphotograph—May 2, 1914.

Analysis— %

Co.....................................97·09
Ni..................................... none
Fe..................................... 1·45
C..................................... 0·067
S..................................... 0·012
Mn..................................... 2·04
Si..................................... 0·011
Ca..................................... none
P..................................... 0·010

Tensile breaking load—75,200 lbs. per square inch.

Etching—strong iodine—9 minutes.

Magnification—130 diameters.

Exposure—2 seconds.

"Commercial Cobalt" H 87c

Tensile strength bar.

Cast from just above the melting point, allowed to cool in iron mould, and turned in lathe. Annealed at 850°C.

Date of microphotograph—May 6, 1914,

Density—8·6658 at 17°C.

Analysis— %

Co.....................................97·8
Ni..................................... 0·50
Fe..................................... 1·46
S..................................... 0·020
C..................................... 0·18
Ca..................................... trace
Si..................................... 0·020
P..................................... trace

Tensile breaking load—60,200 pounds per square inch.
Etching—Strong iodine—10 minutes.
Magnification—130 diameters.
Exposure–1 second.

"Commercial Cobalt" H 87c

Tensile strength bar.
Cast from just above the melting point, allowed to cool in iron mould, and
turned in lathe.
Date of microphotograph—May 6, 1914.

Analysis—

	%
Co.	97·8
Fe.	1·46
Ni.	0·50
C.	0·18
S.	0·020
Si.	0·020
Ca.	none
P.	0·012

Tensile breaking load—56,100 pounds per square inch.
Etching—strong iodine—1 minute.
Magnification—130 diameters.
Exposure—2 seconds.

"Commercial Cobalt" H 109

Tensile strength bar.
Cast from just above the melting point, allowed to cool in iron mould, and
turned in lathe.
Date of microphotograph—May 6, 1914.
Density—8·7997 at 18·5°C.

Analysis—

	%
Co.	96·8
Ni.	0·56
Fe.	2·36
S.	0·022
C.	0·063
P.	0·017

Brinell hardness—107.
Tensile breaking load—52,600 pounds per square inch.
Etching—strong iodine—$\frac{1}{2}$ minute.
Magnification—130 diameters.
Exposure—$\frac{1}{2}$ second.

"Commercial Cobalt" H 211

Tensile strength bar.
Cast from just above the melting point, allowed to cool in iron mould, and
machined in lathe.
Date of microphotograph—May 6, 1914.

Analysis—

	%
C.	0·18
S.	0·080
P.	0·031

48

Brinell hardness—128·2.
Tensile breaking load—31,000 pounds per square inch.
Etching—strong iodine—1 minute.
Magnification—130 diameters.
Exposure—1 second.
This sample shows "ghosts" or "ghost lines" because it is suffering from
   segregation of its impurities, C, S, and P. Metals of this kind are usually
   brittle, weak and hard, which are the characteristics of this particular
   sample as shown under the tables of measurements preceding.

Pure Nickel

Tensile strength bar.
Cast from just above the melting point, allowed to cool in iron mould, and
   turned in lathe.
Date of microphotograph—May 7, 1914.

Analysis—	%
Ni	99·29
Fe	0·48
Co	none
S	0·025
Si	0·042
Ca	none
C	none

Brinell hardness—83·1.
Tensile breaking load—18,000 pounds per square inch.
Melting point—1,452°C.
Etching—nitric acid,—specific gravity 1·42—4 seconds.
Magnification—130 diameters.
Exposure—2 seconds.
This nickel shows polyhedral crystalline structure of the pure metal.

ACKNOWLEDGMENTS.

The analyses throughout the paper were made by Mr. R. C. Wilcox, part-time assistant in the Research Laboratory of Applied Electro-chemistry and Metallurgy, Queen's University; and valuable assistance was rendered by Messrs. W. L. Savell, B.Sc., and K. B. Blake, S.B., both part-time research associates—in conducting certain of the experiments.   The authors wish to acknowledge their indebtedness to these gentlemen.

PLATE XIV.

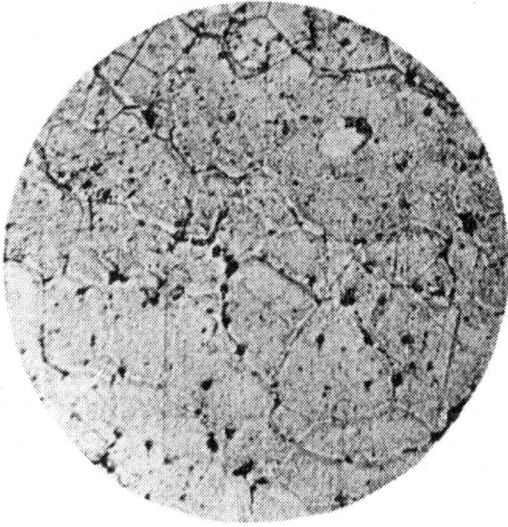

Pure Nickel.

PART II

COBALT

ITS

OCCURRENCE, METALLURGY, USES AND ALLOYS

By

Charles W. Drury

CONTENTS

iii

LIST OF DIAGRAMS

MAP—(*Insert*)

Introductory Note by Author

In preparing this review of the occurrence, metallurgy, uses, chemistry and alloys of cobalt the author has made an attempt to collect all published information concerning this metal, and has added some remarks of his own.

Owing to the growing importance of cobalt and its compounds, and also to the interest that is being taken in the metal because of the similarity of some of its properties to those of nickel, it was felt a compilation and summary of all scattered information should be made. Besides having the general information in accessible form, it is hoped that the various summaries and the list of references in the review, especially those in the section on alloys, will be of assistance to any investigators attempting to study the metallurgy or the alloys of the metal cobalt.

C. W. D.

Queen's University,
  Kingston, Canada,

COBALT

ITS OCCURRENCE, METALLURGY, USES AND ALLOYS

By Charles W. Drury

CHAPTER I

COBALT MINERALS, THEIR COMPOSITION AND OCCURRENCES

Summary of Most Important Deposits of Cobalt

Of the present known deposits of cobalt ores there are only five that are either being worked or are suitable for working at present, viz., those of Cobalt, Ont., Missouri, New Caledonia, Belgian Congo, and Schneeberg, Germany.

Those of Cobalt are the largest, and the shipping ore and concentrate contains an average of 7 to 10 per cent. of cobalt, 5 per cent. of nickel, 25 per cent. of arsenic and 300 to 1,000 ounces of silver. The higher grade silver ores are associated with calcite in narrow veins. The cobalt and nickel minerals occur chiefly as arsenides.

The ores of southeastern Missouri, occurring in the vicinity of Fredericktown, are associated with an entirely different class of minerals, as may be seen from the following approximate composition of the ore: copper, 2 to 3 per cent.; lead, 1.5 to 2.5; cobalt, 0.5 to 0.7, and nickel, 0.7 to 0.9 per cent. Arsenic and silver are not found with the Missouri cobalt ores. The copper and lead occur chiefly as sulphides, and the cobalt and nickel are more closely associated with the copper than the lead minerals. Iron and zinc sulphides are also present.

In the New Caledonia deposits, the cobalt occurs chiefly in the oxide form, the ore averaging 3 to 4 per cent. cobalt oxide. Manganese, nickel and iron oxides occur with the cobalt.

The association of cobalt with copper in Belgian Congo is of importance. As mentioned again in the report, in 1913 8,064 tons of copper, containing from 2.8 to 3.25 per cent. cobalt, were exported to Germany to be refined electrolytically. As this tonnage could yield perhaps 150 tons of cobalt, it is important when the total world's production is not more than about 500 tons.

The cobalt ores of Schneeberg are associated with nickel and bismuth minerals, the bismuth being the most important. Few analyses have been published showing the content of the ore, but those given for development work show about 2 per cent. of cobalt and 1 per cent. of nickel, the percentage of bismuth being considerably more than of cobalt and nickel.

It is a most difficult task to estimate the tonnage of cobalt ores available or developed. There is no question that the Canadian deposits will be the greatest source of cobalt for some years. The ores of Missouri, although containing considerably less cobalt than those of the other countries, are extensive, and will be sufficient to supply practically all the demands in the United States, which are now about 150 tons annually. The Missouri ores are being treated at present by the Missouri Cobalt Company, which is the only producer of cobalt in the United States. The reason why the deposits of New Caledonia are not being operated is the low price of cobalt and the high cost of transportation. However, if the price of cobalt continues to advance, there is no doubt that they will be reopened.

1

Cobalt Minerals

Cobaltite, Cobalt Glance.

Isometric; pyritohedral. Commonly in cubes, or pyritohedrons, or combinations resembling common forms of pyrite. Also granular, massive to compact.

Cleavage; cubic, rather perfect. Fracture uneven. Brittle. H.=5.5. G.=6-6.3. Lustre metallic. Colour; when freshly broken, silver-white inclined to red, also steel-gray with a violet tinge, or grayish black when containing much iron. Streak grayish black.

Composition; sulpharsenide of cobalt, $CoAsS$ or $CoS_2.CoAs_2$, sulphur 19.3, arsenic 45.2, cobalt 35.5.

Occurs at Tunaberg, Riddarhyttan, Vena, and Hakansbö, in Sweden; at the Ko and Bjelke mines of Nordmark; also at Skutterud in Norway. Other localities are Querbach in Silesia; Schladming, Styria; Siegen in Westphalia (from the Hamberg mine, the ferrocobaltite); Dobsina, Hungary; Val d'Annivier, Valais; Botallack mine, near St. Just, Cornwall; Daschkessan, near Elizabethpol, in the Caucasus; Khetri mines, Rajputana, India. The ore from the Khetri mines was sold to the Indian enamellers and jewellers under the name of sehta. Deposits also occur at Tambillos and at Huasco, Chile.

In the United States, it occurs with chalcopyrite and gold in the Quartzburg district, near Prairie City, Grant county, Oregon.

In Canada, at Cobalt, Ontario.

Smaltite, Gray Cobalt Ore.

Isometric; pyritohedral. Commonly massive; in reticulated and other imitative shapes.

Cleavage; o distinct; a in traces. Fracture granular and uneven. Brittle. H.=5.5-6. G.=6.4-6.6 Lustre metallic. Colour; tin-white, inclining when massive, to steel-gray, sometimes iridescent, or grayish from tarnish. Streak grayish black. Opaque.

Composition; essentially cobalt diarsenide. $CoAs_2$, arsenic 71.8, cobalt 28.2.

Occurs usually in veins, accompanying ores of cobalt and of silver and copper. Such associations are found at Freiberg, Annaberg, and particularly Schneeberg, in Saxony; at Joachimsthal in Bohemia, the reticulated varieties are frequently found embedded in calcite; also at Wheal Sparnon in Cornwall; at Tunaberg in Sweden; Allemont in Dauphiné; at the silver mines of Tres Puntas and Veta Blanca, Chile.

In the United States, it occurs in calcite gangue, associated with small quantities of erythrite and native silver, near Gothic, Gunnison county, Colorado.

In Canada, at Cobalt, Ontario.

Linnæite, Siegenite, Cobalt Pyrites.

Isometric. Commonly in octahedrons. Also massive, granular to compact.

Cleavage; cubic, imperfect. Fracture; uneven to subconchoidal. Brittle. H.=5.5. G.=4.8-5. Lustre; metallic. Colour; pale steel-gray, tarnishing copper-red. Streak blackish gray.

Composition; $Co_3S_4=CoS.Co_2S_3$, sulphur 42.1, cobalt 57.9. The cobalt is replaced by nickel (siegenite) and to some extent by iron and copper in varying proportions.

Occurs in gneiss with chalcopyrite, at Bastnaes, near Riddarhyttan, and at Gladhammar, Sweden; at Müsen, near Siegen, in Prussia, with barite and siderite; at Siegen (siegenite), in octahedrons.

In the United States, it occurs with chalcopyrite, pyrrhotite, galena, and bornite in a number of mines of southeastern Missouri, especially in Madison county, at Mine la Motte and Fredericktown, and in St. Francois county; also at Lovelock's Station, Nevada; at Mineral Hill copper mines, Carroll county, Maryland, and at Finksburg, Maryland, associated with copper ores, sphalerite and pyrite, in chlorite schist.

Erythrite, Cobalt Bloom, Red Cobalt, Cobalt Ochre.

Monoclinic. Crystals prismatic and vertically striated. Also in globular and reniform shapes, having a drusy surface and a columnar structure; sometimes stellate. Also pulverulent and earthy, incrusting.

Cleavage: b highly perfect. Sectile. H.=1.5-2.5. G.=2.9. Lustre of b pearly, other faces adamantine to vitreous, also dull and earthy. Colour; crimson and peach-red, sometimes gray. Streak a little paler than the colour, the dry powder deep lavender-blue.

Composition; hydrous cobalt arsenate, $Co_3As_2O_8.8H_2O$. As_2O_5 38.4, CoO 37.5, water 24.1. The cobalt is sometimes replaced by nickel, iron, or calcium.

Occurs at Schneeberg in Saxony, in micaceous scales; in brilliant specimens, consisting of minute aggregated crystals, at Saalfeld in Thuringia; also at Riechelsdorf in Hesse; Modum in Norway; with bismuth at Bieber in Hesse; Andalusia, Spain; Piedmont, Italy. The earthy peach-blossom varieties have been observed at Allemont, in Dauphiné, France; at the Botallack mine, near St. Just, Cornwall; near Alston in Cumberland; and near Killarney in Ireland. A perfectly green variety occurs at Platten in Bohemia, and sometimes red and green tinges have been observed on the same crystal.

In the United States, erythrite occurs in the northeast part of Churchill county, Nevada; near Blackbird, Lemhi county, Idaho; at Josephine mine, Mariposa county, and at Kelsey mine, Los Angeles county, California; also at Lovelock mine, Humboldt county, Nevada.

In Canada, at Cobalt, Ontario.

Willyamite.

Cleavage; cubic. Fracture; uneven. Brittle. H.=5.5. G.=6.87. Lustre; metallic. Colour; between tin-white and steel-gray. Streak; grayish black.

Composition; $CoS_2.NiS_2.CoSb_2.NiSb_2$. A sulph-antimonide of cobalt and nickel.

Found at the Broken Hill mines, in Willyama township, New South Wales, associated with dyscrasite in a calcite and siderite gangue.

Skutterudite.

Isometric; pyritohedral. Also massive granular.

Composition; cobalt arsenide, $CoAs_3$.

Cleavage; a, distinct. Fracture uneven. Brittle. H.=6. G.=6.72-6.86. Lustre; metallic. Colour; between tin-white and pale lead-gray, sometimes iridescent.

Found at Skutterud, near Modum, Norway, in a hornblende gangue in gneiss with titanite and cobaltite, the crystals sometimes implanted on those of cobaltite.

Bismutosmaltite.

Composition; $Co(As.Bi)_3$. A skutterudite containing bismuth. Colour; tin-white. $G.=6.92$.

Occurs with other bismuth minerals at Zschorlau, near Schneeberg, Saxony.

Asbolite, Asbolan, Earthy Cobalt, Wad.

Composition; an impure mixture of manganese and other metallic oxides. Some varieties have been known to contain as much as 32 per cent. of cobalt oxide.

Occurs at Riechelsdorf in Hesse; in Westerwald district between Rhenish Prussia and Westphalia; at Saalfeld in Thuringia; at Nerchinsk in Siberia; at Alderley Edge in Cheshire; Asturias, Spain; and New Caledonia.

An earthy cobalt mineral occurs at Mine la Motte, Missouri, associated with copper, iron, nickel, lead and sulphur; also near Silver Bluff, South Carolina.

Roselite.

Triclinic.

Composition; $(Ca.Co.Mg)_3As_2O_8.2H_2O$. $H.=3.5$. $G.=3.5-3.6$. Occurs in small crystals; often in druses and spherical aggregates. Colour; light to dark rose-red.

Found at Schneeberg, Saxony, on quartz (1842); later obtained from the same region at the Daniel and Rappold mines; also reported from Schapbach, Baden.

Sphærocobaltite.

Rhombohedral. In small spherical masses, with crystalline surfaces, rarely in crystals. $H.=4$. $G.=4.02-4.13$. Lustre; vitreous. Colour; rose-red altering to velvet-black. Streak; peach-blossom red.

Composition; cobalt carbonate—$CoCO_3$.

Occurs sparingly with roselite at Schneeberg, Saxony.

Remingtonite.

A rose-coloured incrustation, soft and earthy. Streak; pale rose-coloured.

Composition; a hydrous cobalt carbonate.

Occurs as a coating on thin veins of serpentine, which traverse hornblende and epidote at a copper mine near Finksburg, Carroll county, Maryland.

Safflorite.

Orthorhombic. Similar to smaltite, essentially cobalt diarsenide, $CoAs_2$. Form near that of arsenopyrite. Usually massive, sometimes showing fibrous radiated structure. Lustre; metallic. Colour; tin-white, soon tarnishing to dark gray.

Occurs with smaltite and implanted upon it, at Schneeberg, Saxony. Also similarly associated at Bieber, near Hanau, in Hesse; at Wittichen in Baden; Tunaberg in Sweden.

Glaucodot.

In orthorhombic crystals. Also massive. Brittle. H.$=$5. G.$=$5.90-6.01. Lustre; metallic. Colour; grayish tin-white. Streak; black.

Composition; a sulpharsenide of cobalt and iron. (Co,Fe)AsS.

Occurs in chlorite slate with cobaltite in the province of Huasco, Chile; also at Hakansbö, Sweden.

Alloclasite.

Commonly in columnar to hemispherical aggregates. H.$=$4.5. G.$=$6.6. Colour; steel-gray. Streak; nearly black.

Composition; probably essentially Co(As,Bi)S, with cobalt in part replaced by iron; or a glaucodot containing bismuth.

Occurs at Orawitza, Hungary.

Bieberite.

Composition; hydrous cobalt sulphate; $CoSO_4.7H_2O$.

Occurs as rose-coloured stalactites in the old mines at Bieber, in Hesse; at Leogang, in Salzburg; at Tres Puntas, near Copiapo, Chile.

Cobaltomenite.

Probably cobalt selenite, $CoSeO_3.2H_2O$.

Occurs with chalcomenite (hydrous cupric selenite), at Cerro de Cacheuta, Argentina.

Jaipurite.

Composition; described as a simple cobalt sulphide, CoS, occurring massive. G.$=$5.45. Colour; steel-gray.

It is stated that this mineral was used by Indian jewellers.

Carrollite.

Isometric, rarely in octahedrons. Usually massive. H.$=$5.5. G.$=$4.85. Colour; light steel-gray, with a faint reddish hue.

Composition; a copper cobalt sulphide, $CuCo_2S_4=CuS.Co_2S_3$, cobalt 38.0, copper 20.5, sulphur 41.5.

Occurs at the Patapsco mine, near Finksburg, Carroll county, Maryland, and also at the Springfield mine, associated and mixed with chalcopyrite and chalcocite.

Sychnodymite.

Isometric, in small steel-gray octahedrons.

Composition; essentially $(Co,Cu)_4S_5$. Part of the cobalt may be replaced by nickel.

Occurs, associated with quartz, siderite, and tetrahedrite, at the Kohlenbach mine, southeast of Eiserfeld in the Siegen district, Prussia.

Pateraite.

An impure, massive mineral of black colour, supposed to be a molybdate of cobalt. Vogl discovered pateraite associated with uranium ores in the Elias mine, Joachimsthal, Bohemia.

Transvaalite.

An oxidation product of cobalt arsenide.

Occurs in black nodular masses forming veins in quartzite at a cobalt deposit, 30 miles north of Middleburg, Transvaal, South Africa. An analysis of a sample of transvaalite showed 1 oz. 12 dwt. of gold per ton.

Heterogenite; Heubachite; Winklerite.

These are hydrated oxidized cobalt minerals, containing nickel, copper, iron, or manganese.

Cobalt is also an occasional constituent of many other minerals, especially of pyrrhotite (sulphide of iron) and arsenopyrite (sulpharsenide of iron), and is usually present in nickel ores. Cobaltiferous varieties of arsenopyrite, known as danaite, are probably due to isomorphous intergrowths of glaucodot.

Metallic cobalt has been recorded as occurring in meteorites.

Summary of the World's Cobalt Deposits[*]

[*]A large part of the description of the world's cobalt deposits as given in this section is taken from the Nineteenth Annual Report of the Ontario Bureau of Mines, Part II, by Willet G. Miller.

Germany and Austria

Deposits of cobalt-silver ores resembling those found in the vicinity of Cobalt, Canada, are found in Germany and Austria. The two areas in these countries are those of Annaberg and Joachimsthal. Mining was begun in the latter about the end of the fifteenth century, while it is stated the deposits of the former were discovered in 1492.

The ores of those two regions are similar to those of Ontario, and contain compounds of cobalt, nickel, bismuth, and silver, with the ore of uranium, which has not been found in the Ontario deposits. The rocks belong to the older systems, but are different in composition from those of Cobalt.

At Joachimsthal, in Bohemia, there is a series of mica schist and limestone cut by dikes of basalt. The veins are said to be older in age than the basalt, and younger than the other rocks mentioned. The veins, which are narrow, contain quartz, hornstone, calcite, and dolomite as gangue material, and often show a brecciated structure.[1] The minerals of these veins are given in the following list:

1. Silver ores: native silver, argentite, polybasite, stephanite, tetrahedrite, proustite, pyrargyrite, and sternbergite.

2. Nickel ores: niccolite, chloanthite, and millerite.

3. Cobalt ores: smaltite as well as bismuth-bearing linnæite and asbolite.

4. Bismuth ores: native bismuth together with bismuthinite and bismuth ochre.

5. Arsenic ores: native arsenic and arsenical pyrites.

6. Uranium ore: uraninite or pitchblende.

[1] Beck, Erzlagerstätten, p. 290, 1903. Beyschlag, Krusch and Vogt, Die Lagerstätten der Nutzbaren Mineralien und Gesteine. Stelzner-Bergeat, Die Erzlagerstätten.

Associated with these are galena, zinc blende, pyrite, marcasite, and copper pyrites.

Among these ores those of cobalt and nickel are generally the older; those of silver the younger. The veins cut through dikes of quartz-porphyry, and are in turn cut by basalt and later dikes.

Of similar composition to those of Joachimsthal are the veins of Annaberg in Saxony. In this neighbourhood the rock is gray gneiss. There are two groups of veins, but the younger, carrying the silver-cobalt ores, are the more important of the ore bodies. The gangue material is chiefly barite, fluorspar, quartz, and breunnerite, with various cobalt, nickel, and bismuth ores, viz., chloanthite, smaltite, red and white nickel pyrites, annabergite, native bismuth, and rarely bismuthinite. Of the silver ores, pyrargyrite, proustite, argentite, native silver, and silver chloride are found. The subordinate gangue minerals are hornstone, chalcedony, amethyst, calcite, aragonite, kaolin, and gypsum; while among the ores are copper pyrites, galena, zinc blende, pyrite, marcasite, tetrahedrite, siderite, uraninite, uranochalcite, uranochre, gummite, and native arsenic. The fact that a large amount of silver chloride was mined at one time, is interesting.

From a large number of observations, the following is given as the relative ages of the various minerals of the Annaberg veins:

V. Decomposition products: annabergite and cobalt bloom.

IV. Silver ores and native arsenic.

III. Calcite and uraninite.

II. Breunnerite and cobalt-nickel-bismuth ores.

I. Barite, fluorspar, and quartz.

The silver-cobalt veins cut the older tin and lead veins of the district as well as the dikes of microgranite and lamprophyre. The latter, especially, are often cut by the silver-cobalt veins. These are cut by basalt, which occurs not only in true dikes, but also in boss-like forms.

Somewhat similar silver-cobalt ores are found in certain veins at Schneeberg, but they are not so strikingly like those of Joachimsthal and Annaberg. A like association of ores is found at Wittichen, where the veins occur in granite.

According to Von Cotta, the rocks of the Joachimsthal district consist of mica schist, together with more or less hornblende schist and crystalline limestone, the whole being cut by numerous dikes of quartz-porphyry and basalt. There are also two large granite masses which rise out of the mica schist. There are lodes of tin, silver, and iron compounds. Tin is found only in the granite region. Silver lodes are divided into four groups fairly distinct from one another. One set contains about 17, while another has 21 lodes. There are also others which do not come to the surface. Von Cotta also states that both classes of lodes intersect the mica schist, with all its subordinate strata, quartz-porphyry, and often even the dikes of basalt and wacké;[1] and that there seem to be cases where dikes of the basalt and wacké have intersected lodes or have penetrated their fissures. From this it may be deduced that the silver lodes were almost contemporaneous with the formation of the basalt, in that their fissures, in part follow the basalt dikes and in part are intersected by the basalt. At all events they stand in a certain genetic connection

---

[1] Wacké is an old term for a soft, earthy variety of trap rock.

to the porphyry, which here is evidently of much greater age than the basalt. The subject is still somewhat obscure. The silver lodes have not yet been found in the granite. Other writers do not agree with Von Cotta, as they appear to be of opinion that the basalt is younger than the veins.

The following notes are taken from Phillips' " Ore Deposits," p. 436, 1896.

The mountains known as Erzgebirge lie on the boundary between Saxony and Bohemia. Joachimsthal lies on the Bohemian side, and is therefore an Austrian town, while Annaberg is in Saxony.

The country rock in the neighborhood of Joachimsthal is for the most part mica schist enclosed between masses of granite. In the eastern part of the mine there are some masses of included limestone, but in the western part, where the veins are not infrequently associated with dikes of porphyry, the gangue is almost quartzose. There are seventeen veins striking north and south, and seventeen others of which the direction is east and west. It has been constantly observed that the former exhibit a tendency to become enriched where they pass through the porphyry or included limestone, while the latter is not similarly affected when they come in contact with these rocks. The ores mined contained silver, cobalt, nickel, bismuth, and uranium. The uranium ores of Joachimsthal became valuable a few years ago, when it was found that uraninite was the chief commercial source of radium.

In 1864, when one shaft had reached a depth of 1,440 feet, a heavy outburst of water, at a temperature of 25°C. and evolving sulphuretted hydrogen, took place, and greatly interfered with underground operations. It was two years before this water could be successfully tubbed off and mining continued.

Cobalt minerals associated with granite are found in several other localities in the Erzgebirge, as well as at Wittichen and at Wolfach, in the Black Forest, where the veins occur in granite.

In Thuringia fault fissures in the Kupferschiefer and Zechstein are filled with barite, calcite and fragments of the country rock, together with smaltite, asbolite, and erythrite. They have been worked especially at Schweina, near Liebenstein.

The palæopicrite of Dillenburg (Nassau) contains cobalt, together with nickel, copper, and bismuth. At Querbach and Giehren, in the Riesengebirge, the mica schist near the contact with gneiss is impregnated with cobaltite, chalcopyrite, pyrite, pyrrhotite, arsenopyrite, blende, galena, magnetite, and cassiterite.

In the Fichtelgebirge, ores of cobalt and nickel are associated with siderite, bismuth, and barite. Siderite and copper ores are found with them in the Siegen district, Prussia.

In Alsace veins of smaltite, chloanthite and native silver in a calcite gangue were formerly worked at Sainte-Marie-aux-Mines.

Messrs. Schmidt and Verloop [1] have described the cobalt and nickel deposits of Schladming, Styria. The predominating orès are given as: niccolite, chloanthite, gersdorffite, and smaltite. Native bismuth and arsenic, arsenical pyrites and löllingite are also found.

---

[1] Schmidt and Verloop, Notiz über die Laerstätte von Kobalt und Nickelerzen bei Schladming in Steiermark: Zeitschr. prakt. Geologie, Vol. XVII, 1910, pp. 271-275.

During the period from 1877 to 1880 there were obtained 29.3 tons of ore, containing 4,497 ounces of silver, 198 pounds of bismuth, 878 pounds of uranic oxide, 1.5 tons of arsenic, and 314 pounds of cobalt-nickel with a little lead, representing a total value of £1,687.

About this time it became evident that the uranic oxide was the most valuable product of these mines, and workings were especially directed to develop the minerals yielding it.

The Schneeberg mines during the year 1881 produced 158 tons of nickel and cobalt ores, valued at £5,902; 1,315 tons of silver ore, and 59 tons of bismuth ore, worth £3,292 and £16,933 respectively.

The production of cobalt ores in Prussia and Bavaria is given below.

PRODUCTION OF COBALT ORES IN PRUSSIA.[1]

Year	Cobalt Ore		Year	Cobalt Ore	
	Metric tons	Value		Metric tons	Value
		$			$
1852	233	16,376	1880	48	2,974
1853	12	5,652	1881	33	2,052
1854	14	6,676	1882	66	3,311
1855	10	4,534	1883	98	4,869
1856	6	3,882	1884	67	2,030
1857	3	1,761	1885	59	1,326
1858	1	767	1886	19	808
1859			1887	11	343
1860	0.3	17	1888	33	992
1861	1	72	1889	503	2,739
1862	1	75	1890	651	10,739
1863	1	291	1891	576	9,209
1864	143	1,470	1892	534	14,550
1865	0.2	37	1893	203	8,491
1866			1894	203	5,741
1867	23	12,313	1895	120	6,298
1868	25	8,371	1896	181	9,868
1869	27	6,764	1897	121	6,256
1870	16	3,762	1898	34	1,700
1871	18	4,228	1899	17	850
1872	219	14,547	1900	4	160
1873	286	13,820	1901	36	2,168
1874	254	35,757	1902	76	
1875	200	19,789	1903	65	
1876	158	19,076	1904	41	
1877	70	5,233	1905	22	
1878	46	2,955	1906	7	
1879	49	3,074			

Cobalt products were made in Prussia continuously from 1852 to 1911. No cobalt ore was produced in Prussia after 1906 and very little since 1897, so that most of the ore treated must have been imported.

In 1884 and 1885 only was there any production of cobalt ores in Bavaria. During these years there were produced 160 tons and 349 tons of ore, valued at $600 and $1,050 respectively.

---

[1] Most of the figures in this and the following tables were taken from the Mineral Industry.

The production of cobalt ore in Saxony has not been published separately, but is given with bismuth and nickel ores. From 1878 to 1901 there was produced 25,350 tons of cobalt, bismuth, and nickel ores.

In 1904, one cobalt-silver mine in the Schneeberg district had a production valued at $132,147. The values were in silver, cobalt, nickel, bismuth, arsenic, uranium, samples, etc. These ores were treated at the "blue colour works," at Schneeberg. Both the government and private companies were interested in the industry. For a detailed description of the cobalt industry in Saxony, the reader is referred to "The Early History of the Cobalt Industry in Saxony," by Mickle, Report of Bureau of Mines, Ontario, Vol. XIX, 1913, pt. ii., pp. 234-251.

PRODUCTION OF COBALT ORES IN AUSTRIA.

Year	Nickel and Cobalt Ore		Year	Nickel and Cobalt Products	
	Metric tons	Value		Metric tons	Value
		$			$
1856	136	18,806	1856		
1857	387	5,508	1857		
1858	342	3,050	1858		
1859	371		1859	11	
1860	281		1860	5	
1861			1861		
1869 (a)	166	662	1869 (a)		
1870	50	108	1870		
1871			1871		
1872			1872		
1873	452	12,244	1873	23	20,150
1874	156	12,546	1874	37	22,460
1875	112	9,864	1875	22	18,892
1876	97	8,340	1876	22	13,734
1877	105	4,790	1877	14	7,896
1878	76	2,242	1878	6	3,502
1879	27	718	1879	5	1,280
1880	16	440	1880	4	1,142
1881	40	200	1881		
1882	15	210	1882	19	1,336
1883	4	158	1883		
1884	5	340	1884		
1885	137	1,546	1885		
1886	37	154	1886		
1887			1887		
1888			1888		
1889			1889		
1890	0.4	126	1890		
1891			1891	1.5	180
1892	0.3		1892	0.15	78
1893			1893	0.12	64
1894	0.5		1894	0.1	62
			1895		
			1896		
			1897	37.6	10,022
			1898	58.8	10,800
			1899	38.1	17,868
			1900	31.2	13,668
			1901	20.5	1,198

(a) From 1861 to 1869 inclusive, no record is given of any production of nickel and cobalt ore or products in Austria.

Deposits of the Chalanches, France

Silver, cobalt, and nickel ores somewhat similar to those of Germany occur in a network of narrow veins in crystalline schist at the Chalanches, in the Dauphiné, France. These deposits were discovered in 1761, and have been described by T. A. Rickard.[1]

The following notes are from Mr. Rickard's paper:

During the earliest period of mining at the Chalanches, some bodies of rich silver, nickel, and cobalt ore were found near the surface. It is said that two shots produced sufficient silver to pay for the two buildings known as the pavilions of Allemont, with their various ornamentations, including the fleur-de-lis which still adorn the roof. As 200 to 300 kilos of silver would at that time be worth from $10,000 to $15,000 this statement does not seem incredible.

It is remarkable that although the silver is always associated in the lodes with rich nickel and cobalt ores, often with bunches of stibnite, and more rarely and erratically with gold, the government engineers took no notice of any metal other than silver. The speiss[2] containing nickel and cobalt was rejected with the slags, and was used to fill the swamps or form road beds, which, in later times, were furrowed and turned over to recover their valuable contents.

The possibility of utilizing three metals instead of one seems to have dawned upon the engineers quite as a discovery; and this fact stimulated the repeated spasmodic attempts to rehabilitate the old mine. The arsenides of nickel and cobalt were sold in England and Germany. A German chemist was employed at Allemont to manufacture cobalt pigments for the arts, but was not successful, and the attempt was abandoned.

In 1891 gold was first recognized. However, its importance proved greater from a scientific than from a commercial point of view. The old mine-workings, aggregating 20 kilometres in length, showed that a great deal of unsuccessful exploration had been carried out.

The geological formation is simple. A network of veins traverses crystalline schists of variable character. The country rock forms a part of the great crystalline formation usually referred to as the Archaic schists of the Alps, though in point of fact they probably include rocks from the Archaic to the Carboniferous. Lithologically, certain sections suggest the Huronian and Laurentian. These schists lie immediately upon the granite; they are extremely variable in character, so that at different places they can be described as gneissoid, granitoid, talcose, micaceous, graphitic, or amphibolic. There are also blocks of rock containing epidote.

The maps of the mine exhibit a wonderful network of galleries spreading like a cobweb over an area of about 600 by 300 metres. It is computed that the workings aggregate in length not less than twelve miles, an extent in remarkable contrast to the relatively small quantity of ore produced.

It has been thought by several observers that the lodes were more numerous near the surface than in the interior of the mine. This is due to the fact that any single fissure, in approaching the surface, spreads itself out into a number of subordinate fractures. It has also appeared that the lodes gained in regularity as they penetrated the mountain. Caillaux, therefore, adds that this fact seems to indicate the probable occurrence in depth of only a small number of lodes, but that those surviving will have a regularity of structure greater than those which have been hitherto exploited. The veins vary in width from a knife-blade to 80 centimetres (31.5 inches); their usual thickness lies between 3 to 30 centimetres (0.1 to 1 foot).

Examination of the old workings proves clearly that farther from the surface the country rock gets harder, the vein matter loses its soft character, and the veins become fewer in number, more regular, narrower, and less ore-bearing. Approaching the surface, on the contrary, the schists are fractured in a multiplicity of directions, the veins become larger, their filling is generally earthy, and they throw off branches, at the intersections of which the ore bodies are found. In general, mineralization becomes more pronounced near the surface; this being due, not merely to the oxidation of the sulphides, but to an actual relative increase of 'orey' matter.

---

[1] Rickard, The Mines of the Chalanches, France. Am. Inst. Min. Eng. Trans., Vol. XXIV, 1894, pp. 689-705.

[2] Speiss is a product formed in smelting which contains most of the metals in the form of arsenides.

The observations led to the conclusion that the richest part of the mine was that which was within the influence of oxidation, and that both chemical agencies and structural conditions favoured an enrichment of ore near the surface. This statement is particularly applicable to the silver contents. It also holds true of the gold, but it is less accurate with respect to the nickel and cobalt. The richness in silver of the oxidized ores suggests secondary precipitation. This is confirmed by the fact that the silver appears to be thrown down upon the nickel and cobalt arsenides and often envelops them in such a way as to impart the rudiments of a nodular structure. The hard, undecomposed arsenides contain small amounts of silver. The gold, occasionally present, is associated with soft, maroon-coloured, earthy iron-bearing vein matter. The nickel and cobalt minerals appear to be primary, and are more persistent than those of silver and gold.

If we accept the current theory that the nickel and cobalt came from the leaching of magnesium silicates (and facts are numerous pointing that way), then we must conclude that the origin of the nickel and cobalt ores of the Chalanches was not the immediately adjacent rocks, but ones similar, which underlie them at a greater depth. The silver and gold, it may be suggested, were precipitated from other solutions, and at a period other than that which saw the deposition of the nickel and cobalt. The precious metals were probably derived from a deeper-seated source, and may have been leached from the granite which underlies the schists and is penetrated by the basic eruptives. In both cases the various metals must have come from a depth where leaching action was powerful, and from which ascending currents brought the metallic constituents, the subsequent precipitation of which produced valuable ore-deposits.

Quartzose veins containing ferriferous smaltite were prospected in 1784 at Juzet, near Nance, Montauban-de-Luchon, Haute-Garonne. The ores produced together with those from Gistain on the Spanish side of the Pyrenees, were treated at Saint-Marnet.

Norway

The following description of the cobalt deposits of Norway is given by Phillips:[1]

The cobaltiferous fahlbands of the districts around Skutterud and Snarum occur in crystalline rocks varying in character between gneiss and mica schist; however, from the presence of hornblende, they sometimes pass into hornblende schists. These schists, of which the strike is north and south, and which have an almost perpendicular dip, contain fahlbands very similar in character to those of Kongsberg. They differ from those of that locality, however, inasmuch as while here the fahlbands are often impregnated with ore, those of Kongsberg, although to some extent containing disseminated sulphides, are only of importance as being zones of enrichment for ores which occur in veins. The ore zones usually follow the strike and dip of the surrounding rocks, and vary in width from 15 to 30 feet. The distribution of the ores is by no means equal. The predominant rock of the fahlbands is a quartzose, granular mica-schist, which gradually passes into quartzite, ordinary mica-schist, or gneiss. The ores were cobalt glance, arsenical and iron pyrites, molybdenite, and galena. It is remarkable that in these mines nickel ores do not accompany the ores of cobalt in any appreciable quantity. The principal fahlband is known to extend for a distance of about six miles, and is bounded on the east by a mass of diorite which protrudes into the fahlband, while extending from the diorite are small dikes or branches traversing it in a zigzag course. It is also intersected by dikes of coarse-grained granite which do not contain any ore, but which penetrate the diorite.

The Skutterud mine in 1879 produced 7,700 tons of cobalt ore, which yielded 108 tons of cobalt schlich,[2] containing from 10 to 11 per cent. of cobalt, and worth about £11,000. These deposits, which at one time were among the world's chief producers of cobalt, are of too low grade to be worked now.

Silver veins were discovered at Kongsberg in 1623. These veins resemble those of Cobalt in width, in that the numerous veins or stringers occur in small areas, and in the gangue mineral which is essentially calcite. However, nickel and cobalt minerals are not characteristic of the Norwegian deposits.

---

[1] Phillips, Ore Deposits, 1884, p. 390.
[2] Schlich is a term given to washed cobalt ore.

PRODUCTION OF COBALT ORES IN NORWAY.

Year	Cobalt Ore		Year	Cobalt Ore	
	tons	value		tons	value
		$			$
1866	85	20,000	1884	90	12,922
1867	80	7,500	1885	101	13,702
1868	5	7,568	1886	123	11,960
1869	80	14,456	1887	57	4,160
1870	10	4,056	1888	84	8,060
1871	25	18,520	1889	152	14,300
1872	65	65,152	1890	213	19,500
1873	75	40,560	1891	187	13,000
1874	65	45,100	1892	123	8,580
1875	60	52,500	1893	123	12,150
1876	95	45,400	1894	89	8,100
1877	105	52,000	1895	45	4,050
1878	105	52,000	1896	29	2,700
1879	108	52,000	1897	24	2,700
1880	87	45,500	1898	21	2,160
1881	80	41,000	1899	...	...
1882	99	52,000	1900	...	...
1883	84	15,556			

Most of the cobalt ore mined in Norway was exported.

In 1896 the cobalt works at Modum operated on a limited scale, employing 36 men.

Sweden

At Tunaberg, cobalt ore occurs in a bed of granular limestone in gray gneiss. The limestone contains, principally, hornblende, mica, and serpentine; also lead, silver, copper and cobalt minerals, copper pyrites and cobalt glance being the most frequent.

Other localities are Vena, near Askersund, on Lake Wetter, and at Gladhammar, south of Westerwik.

The following table shows the extent of the working of the cobalt deposits in Sweden.

PRODUCTION OF COBALT ORES IN SWEDEN.

Year	Cobalt Ore	Year	Cobalt Ore
	Metric tons		Metric tons
1870	58	1882	516
1871	....	1883	177
1872	41	1884	56
1873	17	1885	137
1874	41	1886	164
1875	74	1887	231
1876	101	1888	143
1877	153	1889	266
1878	726	1890	145
1879	223	1891	244
1880	331	1892	53
1881	556	1893	101

In the years 1887, 1888 and 1889 there were also produced 376, 258, and 177 kilos of concentrated ore.

In 1888, 1889, 1890, 1891, 1892, and 1893, there were produced 7,270, 14,154, 15,444, 13,772, 15,703, and 7,255 pounds of cobalt oxide respectively.

Italy

Cobalt and nickel ores occur with quartz, calcite, and ores of copper in Piedmont.

Switzerland

Cobalt and nickel ores occur in Valais, and at Ayer in the Val d'Annivier and at Kaltenberg in Turktmanntal.

New Caledonia

Until the discovery of the cobalt deposits, at Cobalt, Ont., Canada, in 1903, about six countries were supplying the world with cobalt ores, New Caledonia producing probably 85 or 90 per cent.

When the ore from Ontario was put on the market, the prices fell materially in New Caledonia, and cobalt mining has now practically ceased in that country. It seems strange that Ontario should be the only serious competitor which this French colony, in the Southern Pacific, has in both nickel and cobalt.

The cobalt deposits of New Caledonia occur under conditions similar to those of nickel, and the two metals are frequently associated in economic quantities. New Caledonia is a non-glaciated country. The rock peridotite underlies a considerable part of the surface. This rock weathers readily, and so, over a large part of the surface, the alteration product, serpentine, occurs. The surface of the serpentine is more or less broken down, forming comparatively loose or slightly coherent deposits. It is in association with these that the cobalt is found, as asbolite, earthy cobalt, or cobaltiferous wad. Asbolite is a mixture of oxides of cobalt, manganese, and other metals and can hardly be called a distinct mineral. It has been proved that the cobalt, nickel, and other metals found in this decomposed rock were originally constituents of the peridotite.

The peridotites are believed by some writers to be post-Cretaceous in age, and are said to be in the form of a surface flow covering the uneven or eroded surface of the underlying Cretaceous strata. These rocks which are high in magnesia and low in iron, constitute the great serpentine formation of New Caledonia. They are more or less charged, when fresh, with crystals of ferro-magnesian pyroxene. The unaltered rock belongs, therefore, in Rosenbusch's classification, to harzburgite. Dunite, a variety of peridotite of which the main constituent is olivine, is found with chrome iron ore. The peridotites usually show traces of advanced alteration, which has resulted in more or less transformation of olivine to serpentine, and in the development of talc from pyroxene. At times the alteration is sufficiently advanced to produce perfect crystals of antigorite, with some films of talc.

Since these rocks always contain a little manganese, nickel, and cobalt, it would appear that these metals occur in the olivine as well as in the enstatite. Grains of chrome ore are abundant in all samples. The rocks are often traversed by less basic dikes of the character of gabbro. Diorite sometimes outcrops in the middle of the serpentine exposures.

In one deposit,[1] it is said that the decomposed material occupies a profound depression in the serpentine. This basin is filled by a red, clay-like deposit which has a depth of about 52 metres in the centre and 10 or 12 metres around the border. The richest ores appear to occur near the centre of the basin and near the contact of the serpentine.

It will be seen that all the cobalt deposits are irregular in form, and hence it is difficult to estimate their value.

Much of the mineral mined contains only two or three per cent. of oxide of cobalt and after washing probably 4.5 per cent. Cobalt ores from New Caledonia were purchased (1904) on a basis of 4 per cent. CoO, for which ore 330 francs [2] ($66.00) per ton was paid. In purchasing ores containing cobalt in excess of 4 per cent. CoO, a premium, which varied with the grade of ore, was allowed. The grades were divided as follows: those containing between 4 and 5 per cent. cobalt oxide. 5 and 6 per cent., 7 and 8 per cent., and above 8 per cent. Premiums of 0.8, 0.9, 1.0, and 1.5 francs for the different grades respectively were allowed for each 0.1 per cent. above the minimum per cent. of each grade. On this basis mineral carrying 8 per cent. would be worth 750 fr. ($150) a ton.

The prices paid for cobalt ore in New Caledonia in 1908 were about as follows: for 4.5 per cent. ore, $23 a ton; for 5 per cent. ore, $27 to $28 and 90 cents for each 0.1 per cent. above. At these prices, only rich, well-situated, and developed mines could be worked. An 8 per cent. New Caledonia ore at the price quoted would bring $54 a metric ton.

Ouvrard [3] reports the price of cobalt ore from New Caledonia and Chile as $35.00 per ton, c.i.f. European ports, for ore containing 4 per cent. cobalt; with a premium or penalty of $1.20 for each 0.1 per cent. cobalt above or below 4 per cent.

Previous to 1910 the Anglo-French Nickel Company of Swansea, Wales, bought some ore at Cobalt for the cobalt content alone, and paid 30 cents a pound for the metallic cobalt in an 8 per cent. ore. This is about $53 per ton for such ore.

The following analysis shows the composition of the New Caledonia ores: MnO_2 18, CoO 3, NiO 1.25, Fe_2O_3 30, Al_2O_3 5, CaO and MgO 2, silica 8, and loss on ignition 32 per cent.

In the following table the amount of cobalt ore exported from New Caledonia between 1893 and 1909 is given.

---

[1] Glasser, Report to the Minister of the Colonies on the Mineral Wealth of New Caledonia, 1904.

[2] A franc is worth about 20 cents.

[3] Ouvrard, Industries du Chrome, du Manganese, du Nickel, et du Cobalt, p. 253, Doin and Sons, Paris, 1910.

EXPORTATION OF COBALT ORE FROM NEW CALEDONIA.

Year	Cobalt Ore	Year	Cobalt Ore
	Metric tons		Metric tons
1893....................	(a) 520	1903....................	8,292
1894....................	(a) 4,156	1904....................	8,961
1895....................	5,302	1905....................	7,9.9
1896....................	4,823	1906....................	2,48/
1897....................	4,757	1907....................	3,943
1898....................	2,373	1908....................	3,405
1899....................	3,294	1909....................	979
1900....................	2,438	1910....................	....
1901....................	3,123	1911....................	....
1902....................	(b)	1914....................	(c) 920

(a) In 1893 and 1894, 169 and 7 tons, respectively, of matte were exported.
(b) Not reported.
(c) During 1914, New Caledonia exported 920 tons of ore and 25 tons of matte; Mineral Industry, Vol. 23, 1914, p. 548.

New South Wales

New South Wales was the second largest producer of cobalt ore before the discovery of the deposits at Cobalt, Ontario. The deposits, which are situated near Port Macquarie, are similar to those of New Caledonia.

During 1898, 1899, and 1900, 119, 193, and 145 tons of cobalt ore, valued at $2,800, $4,595, and $7,950 respectively were shipped.

In 1903 the quantity of cobalt ore exported from the deposits near Port Macquarie amounted to 153 tons, valued at $7,850. Since 1903, there have not been any shipments of cobalt ore made from New South Wales.

South Australia

Cobalt ore, containing smaltite and other minerals, occurs at Bimbowrie, near Olary, on the Broken Hill line, but little work has been done on the deposit.

Africa

While, as we have seen, silver has been worked in association with cobalt, the latter metal has been seldom found in association with gold in important quantities. However, one such occurrence is in the Middleburg district in northern Transvaal. In the vein in this district, the gangue material is kaolin, with which is mixed gold-bearing quartz. In the quartz are small nest-like aggregations of smaltite and copper ores, and at times molybdenite, also the secondary minerals cobalt bloom, limonite and skorodite.

A small amount of ore was produced in 1890 and also in 1895. During 1895 a quantity of ore analyzing 7 to 10 per cent. was reported to have been exposed. A brief account[1] of one of the cobalt deposits states:

[1] Geology of the Neighborhood of Middleburg, Transvaal Mines Dept., Pretoria, 1907.

Cobalt, in the form of smaltite and erythrite, is found at Balmoral, and also just beyond the northern boundary of the map in the valley of the Kruis river. At Balmoral the cobalt is associated with feldspar and actinolite, together with secondary quartz and calcite, in veins most probably of igneous origin, which traverse a series of highly altered sedimentary rocks of shaly character in the neighbourhood of the junction of the Waterberg and Transvaal systems.

Three additional references to cobalt deposits of the Transvaal are given below.

Beck, Note on Cobalt Lodes of the Transvaal: Trans. Geol. Soc., South Africa, vol. 10, 1907, p. 10.

Mellor, Note on the Field Relations of the Transvaal Cobalt Lodes: Trans. Geol. Soc., South Africa, vol. 10, 1907, p. 36.

McGhie and Clark, Transvaalite—A new Cobalt Mineral from the Transvaal: Jour. Soc. Chem. Ind., vol. 9, 1890, p. 587.

The crude copper produced by the Union Minière du Haut Katanga, Belgian Congo, of which 8,064 tons were produced and shipped to Germany in 1913, contained from 2.8–3.25 per cent. of cobalt. This formed a by-product easily saved in electrolytic refining and has been one of the chief sources of German cobalt in recent years.[1]

Cobalt and nickel oxides in small quantities (less than one per cent.) associated with chromite ores are found in Sekukuneland and in the neighbourhood of Selukiva (Rhodesia).

India

Cobalt ores[2] in small quantities were found in some of the mines of Rajputana, and were used for colouring glass bangles. Ores of this metal also occur in Tenasserim. Both cobalt and nickel are present in small quantities in the pyrrhotite from the Khetri mines, and traces of nickel sometimes occur in iron ores from Bhangarh.

Linnæite has recently been identified among some copper ores from Sikkim.

United States

Although there are no extensive deposits of cobalt in the United States, reports show that small quantities of cobalt oxide have been produced annually since 1869. Most of this was recovered in the refining of the copper-nickel ores of Sudbury, but at present practically all the cobalt is slagged in the converter, so that very little reaches the refinery. There has also been a small amount of cobalt oxide recovered from the lead-copper ores of Missouri, but with the exception of the years 1903 and 1908, the amount obtained has been small. In the years mentioned 120,000 and 100,000 pounds of oxide respectively were produced. Since the discovery of the cobalt deposits at Cobalt, Canada, cobalt ores have been shipped to the United States for treatment. Also the unrefined cobalt and nickel oxides produced at the smelter of the Canadian Copper Company[3] between 1905 and 1913, were shipped to the

---

[1] Min. Sci. Press, Vol. CVIII, 1914, p. 822; Mineral Resources of United States, U.S. Geol. Sur., Part I, 1913, p. 339.

[2] Phillips, Ore Deposits, 1884, p. 436.

[3] The production of the Canadian Copper Co. is given under the Metallurgy of Cobalt.

United States to be separated and purified. Previous to 1900, the Mine la Motte Company shipped ores to Swansea, but about this time a plant was erected at the mine to recover the cobalt oxide. This plant produced 120,000 pounds in 1903, but was closed shortly afterwards. In 1906 the North American Lead Company erected a refinery at Fredericktown, Missouri, to recover cobalt oxide from Missouri ores. This plant was operated during 1907, 1908, and 1909, but was closed in 1910. The reconstruction of this refinery was commenced in 1916 by the Missouri Cobalt Company, and during 1918 a quantity of cobalt oxide was produced. The Missouri Cobalt Company has erected a mill with a daily capacity of 300 tons.

In 1903 cobalt and nickel ore associated with fluorspar was said to have been discovered near Marion, Kentucky.

In 1905 one or two small shipments of cobalt ore from deposits in Grant county, Oregon, were made to France. These deposits are described as occupying fissures in a dark-greenish, partly altered, diabase-porphyry. The ore bodies appear to be more or less lenticular in shape, and vary from a few inches to several feet in width. The principal minerals are chalcopyrite, smaltite, arsenopyrite, pyrite, pyrrhotite, malachite, and bornite with a quartz and calcite gangue. The chief metals present were gold, cobalt, and copper. From a sample of the ore carrying smaltite and chalcopyrite the former mineral was found to have the composition given below, No. 1. This smaltite has a rather unusual appearance, resembling somewhat acicular or fine columnar stibnite. In composition it is close to that from Gunnison county, Colorado, an analysis of which is given by Dana, No. 2 below. The Standard Consolidated Mines Company, Oregon, worked during 1905 a few veins of cobalt ore carrying gold and silver. An analysis of a picked specimen of the ore follows, No. 3.

—	No. 1 [1]	No. 2	No. 3 [2]
Cobalt	14.88	11.59	9.91
Nickel	1.12	trace	0.57
Arsenic	64.06	63.82	42.66
Sulphur	0.57	1.55	
Iron	11.11	15.99	14.93
Insoluble	2.22		
Calcium carbonate	6.34		6.8
Silver			5.2 oz.
Gold			1.62 oz.

Smaltite occurs in a calcite vein in granite at Gothic, Colorado.

Near Blackbird, Lemhi county, Idaho, lenticular bodies of cobalt-nickel ore occur in pre-Cambrian schists and quartzites which are cut by diabase and lamprophyre dikes.

In Los Angeles county, California, cobalt silver ores are found in barytic lodes.

---

[1] Analysis made by A. G. Burrows, Ontario Bureau of Mines.

[2] Mineral Industry, Vol. XIV, 1905, p. 461.

The following table shows the cobalt oxide produced and imported into the United States:

PRODUCTION AND IMPORTS OF COBALT OXIDE.[1]

Year (a)	Production pounds	Imports pounds	Year (a)	Production pounds	Imports pounds
1869	811	....	1894	6,763	24,020
1870	3,854	....	1895	6,400	36,155
1871	5,086	....	1896	12,825	27,189
1872	5,749	....	1897	19,300	24,771
1873	5,128	1,480	1898	9,640	33,731
1874	4,145	1,404	1899	10,200	46,791
1875	3,441	678	1900	12,270	54,073
1876	5,162	4,440	1901	13,360	71,969
1877	7,328	19,752	1902	20,870	79,984
1878	4,508	2,860	1903	120,000	73,350
1879	4,376	7,531	1904	22,000	42,353
1880	7,251	9,819	1905	(b)	70,048
1881	8,280	21,844	1906	....	41,084
1882	11,653	17,758	1907	....	42,794
1883	1,096	13,067	1908	100,000	1,550
1884	2,000	25,963	1909	....	9,818
1885	8,423	16,162	1910	....	6,124
1886	8,689	19,366	1911	....	22,934
1887	5,769	26,882	1912	....	31,848
1888	7,491	27,446	1913	....	28,729
1889	12,955	41,455	1914	....	109,484
1890	6,788	33,338	1915	....	190,145
1891	7,200	25,483	1916	....	238,934
1892	7,869	32,833	1917	....	236,822
1893	8,422	28,164			

[1] Mineral Industry.

(a) Production is stated for calendar years; imports for fiscal years ending June 30 until 1887, and for calendar years subsequently.

(b) Since 1904, with the exception of 1908, no record is given of any cobalt oxide having been produced in the United States. However, since 1905, the total production of mixed cobalt and nickel oxides of the cobalt smelter of the Canadian Copper Co. was shipped to the United States to be refined. The amount recovered from this source may be closely approximated by referring to the production of the Canadian Copper Company. See section entitled " Metallurgy of Cobalt."

The annual consumption of cobalt oxide in the United States amounts to approximately 200,000 pounds, and although the sum of the figures in the table does not equal this amount, except in a few years, this is merely because the cobalt materials imported were classified as some product other than cobalt oxide, e.g. cobalt ore, and zaffer. Zaffer is a term applied to finely-ground roasted cobalt ores or products.

In 1913 the import duty on cobalt oxide was reduced from 25 to 10 cents per pound.

Mexico

Cobalt-bearing minerals have been found at several localities in Mexico, but little has been published concerning these occurrences. Near the village of Pihuamo in the state of Jalisco, cobalt minerals occur in veinlets cutting a large vein of magnetite associated with pyrite and pyrrhotite. The chief rock in the vicinity is described as andesite. It is said that a number of tons of ore containing 8 or 9 per cent. of cobalt were mined in this district. The minerals are cobaltite and small quantities of smaltite and cobalt bloom. The vein matter consists of greenish calcite, and a little barite. Niccolite also appears to be present.

The following Mexican localities are also reported to contain cobalt minerals: Iturbide, in Chihuahua; Guanacevi, in Durango; Cosala, in Sinaloa; at the Mirador mine in Jalisco. Small shipments of ore containing 30 per cent. of cobalt and 7 per cent. of nickel were made from the Esmeralda and Pihuano mines in Jalisco. A pink cobaltiferous variety of smithsonite occurs at Bolco, Lower California. An article dealing with the cobalt deposits in Jalisco has been published by Navarro, Mem. y Rev. Soc. Cientif., Antonio Alzate, vol. 25, 1907, pp. 51-57.

Peru

Nickel and cobalt minerals are reported to occur in the Department of Cuzco.

Chile

Nickel and cobalt minerals are found associated with the ores of silver at several places in Chile. Among these are Mina Blanca de San Juan, Department of Freivna, Province of Atacama; Minillas, Cambillos Brutre, in the Province of Coquimbo; and Tajon del Yeso, in the Province of Santiago. The ores of the Colorado mine of Chanaicillo are nickel-bearing. Veins of nickel ore are found also at San Pedro, near Flamenco, a small port south of Chanaial.

During the latter part of 1901 shipments of cobalt ores were made from Caldera to the United Kingdom, and it was anticipated then that the returns would show sufficient profit to render the re-opening of the mines advisable.

The San Juan group of mines, lying north of the Port of Peña Blanca are located on well formed lodes varying from one to six feet in width. The ore consists of oxide, arsenate, and sulpharsenide of cobalt with an average content of about 4 per cent. of cobalt. The workings reached a depth of 120 to 200 feet.

Cobalt mines[1] were operated in the Provinces of Atacama, Coquimbo, and Aconcagua. The production in 1903 was 284 metric tons of 7.15 per cent. ore. The largest producer was the Rosa Amelia mine, situated in the Department of Freivna, Atacama, this mine in 1903 producing 133 tons of ore. In the Goyenechea mine, in the Department of Chanaial, an argentiferous cobalt ore was mined carrying 8 per cent. of cobalt.

Mining of cobalt glance and erythrite was also carried on at Tambillos and Huasco.

The quantity of cobalt ores mined in Chile between 1844 and 1905 was 6,384 tons.

Argentina

A cobalt deposit was discovered in 1904 at Valla Hermoso, Vinchina, Provincia de la Rioja, Argentina, and was worked on a small scale.[2] The ore occurs on the western slope of the Cerro de Famantina, a spur of the Andes, in a talcose schist, usually near its contact with an acid, igneous rock. A number of veins appear at the surface, but only one has been exploited. The ore body varies in width from 0.9 m. to 1.3 m., with an average of about 1.1 m. The ore consists of cobaltite and

[1] Mineral Industry, Vol. XIII, 1904, p. 338.
[2] Mineral Industry, Vol. XIII, 1904, p. 336.

arsenopyrite in a gangue of quartz. This was hand-cobbed into first and second class ores. The first class assayed 6.0 to 7.0 per cent., and the second class, 3.0 to 4.5 per cent. cobalt. The first class contained 0.75 to 1.0 oz of gold and 5 to 10 oz. of silver per ton. About 300 tons were produced, of which 150 were hand-cobbed, and of this only a few tons were first class ore. The distance of this property from Monozasta, which is the nearest railway station on the F. C. A. Del Norte line, is 120 miles. All the ore that was concentrated was shipped to England.

Great Britain

For many years small supplies of cobalt ore were obtained from the mines at Moel Hiraddug, near Rhyl; Cornwall; in Flintshire, and Cumberland. However, no production is recorded since 1890.

From 1860 to 1890 there were produced 1,242 tons of cobalt and nickel ore valued at $36,710.

Spain

Mining of cobalt ores was carried on in Spain between 1871 and 1897 in the Valley of Gistain, Huesca, near the French frontier. At Guadalcanal, in Andalusia, veins containing ores of silver, cobalt, and sometimes copper, in a calcite gangue, were at one time important. However, only small quantities were produced, as may be seen from the following table:

PRODUCTION OF COBALT ORES IN SPAIN.

Year	Cobalt Ore		Year	Cobalt Ore	
	tons	value		tons	value
		$			$
1871	4		1887	436	13,914
1872	40		1888	68	5,476
1873	4		1889	141	4,876
1874	82		1890	111	4,368
1875	89		1891	60	1,804
1876	115		1892	24	72
1877	433		1893	18	194
1878	100		1894	52	624
1879	110		1895	7	84
1880	129		1896	18	1,800
1881	102	13,720	1897	13	3,400
1882	40	5,234	1898		
1883	19.4	2,486	1899		
1884			1900		
1885			1901		
1886	132	17,200			

Russia

A deposit of cobalt ore, free from nickel, occurs at Dachkessan, Government of Elizabethpol. This deposit is in the form of a dike of diorite impregnated with cobalt minerals associated with iron and copper pyrites. The workings are now abandoned.

China

Very little is known of any cobalt deposits in China. Bowler[1] states that cobalt ore is found at the base of a range of sandstone hills near the town of Tsangscheng, the geological formation being probably Upper Cambrian. These deposits are now either exhausted or very little worked.

Further information about the treatment of these ores will be found under the "Metallurgy of Cobalt."

Ontario, Canada[2]

Situation and Discovery

The ore bodies at Cobalt, which carry silver, cobalt, nickel, and arsenic, were discovered in 1903 during the building of the Timiskaming and Northern Ontario railway. The first of these deposits to be worked lies within half a mile of Cobalt station, which is about 330 miles north of Toronto. One of the oldest known ore bodies in North America, the argentiferous galena on the east side of Lake Timiskaming, is distant only eight miles from Cobalt station. This galena deposit, known as the Wright mine, was apparently discovered by voyageurs over 150 years ago.

It may be added that the building of the Canadian Pacific railway exposed the Sudbury nickel deposits 90 miles southwest of Cobalt. It can thus be said that each of the two railways in this part of Ontario, brought to light an important mineral field.

The Sudbury deposits have received a great deal of attention from geologists of this and other continents. One group, among whom may be mentioned Barlow,[3] Coleman,[4] and Vogt, regard them as due to magmatic segregation from an original igneous magma, without further concentration. Another group, among whom are Dickson,[5] Beck,[6] and Knight,[7] contend that these deposits are the result of aqueous igneous action, and that the sulphides were deposited after the accompanying rocks were formed.

Before proceeding to a consideration of the cobalt veins, a table showing the age relations of the rocks at Cobalt is given below.

---

[1] Bowler, Chinese Treatment of Cobalt Ores. Chemical News, Vol. LVIII, 1888, p. 100.

[2] Miller, Willet G., The Cobalt-Nickel Arsenides and Silver Deposits of Timiskaming, Ontario. Reports of Bureau of Mines, Ontario: Vol. XIV, Pt. II, first edition, 1905; second edition, 1906; third edition, 1908; fourth edition, Vol. XIX, Pt. II, 1913. The larger part of the description of the deposits at Cobalt is taken from Dr. Miller's reports.

[3] Barlow, Nickel and Copper Deposits of the Sudbury Mining District. Geol. Survey of Can., 1901, Pt. H.

[4] Coleman, The Sudbury Nickel Region, Ontario Bureau of Mines, Vol. XIV, 1905, Pt. III; The Nickel Industry, with Special Reference to the Sudbury Region, Ontario, Department of Mines, Ottawa, 1913.

[5] Dickson, The Ore Deposits of Sudbury, Ontario. Am. Inst. Min. Eng. Trans., Vol. XXXIV, 1904, pp. 3-67.

[6] Beck, The Nature of Ore Deposits, p. 41.

[7] Knight, Origin of Sudbury Nickel-Copper Deposits. Eng. and Min. Jour., Vol. CI, 1916, p. 811; also Report Royal Ontario Nickel Commission, 1917.

Age Relations of Rocks of Cobalt and Adjacent Areas [1]

PALEOZOIC
Silurian
Niagara

(*Great unconformity*)

EOZOIC OR PRE-CAMBRIAN
Later Dikes

Nipissing Diabase
(*Intrusive contact.*)

Cobalt Series
(*Unconformity*)

Lorrain Granite
(*Intrusive contact.*)

Lamprophyre Dikes
(*Intrusive contact.*)

Timiskaming Series
(*Unconformity*)

Keewatin Complex

Character and Origin of Cobalt Veins

The deposits at Cobalt occupy narrow, practically vertical fissures, and joint-planes in the metamorphosed Cobalt series. A few productive veins of similar form have been found in the intrusive Nipissing diabase. Others occur in the Keewatin, which is the oldest series of the area, and which consists of basic volcanic rocks. The most productive veins in the Keewatin have been No. 26 on the Nipissing and the vein system on the Timiskaming-Beaver. The former vein lies close to the western edge of the diabase sill. Before erosion of the sill took place, vein No. 26 lay beneath the sill or in its foot-wall. The Timiskaming-Beaver veins, on the other hand, lie in the upper or hanging wall of the sill. There are veins which run from the conglomerate and other fragmental rocks of the Cobalt series into the underlying Keewatin; and there are veins, e.g. the Nova Scotia and Timiskaming veins, which run downward from the Keewatin into the underlying, intrusive Nipissing diabase. A vein on the Cobalt Central passes from the surface downward through the Nipissing diabase into the Cobalt series, which here forms the foot-wall of the diabase sill. Moreover, " blind " veins, or veins that do not outcrop at the surface, have been worked on several properties. One of the most interesting of these occurs beneath Peterson lake. This vein is in the Keewatin, which is here overlain by the Nipissing diabase sill. The vein runs up to the bottom of the sill, but not into it. The figure on page 24 shows the relationship of the rocks and the type veins described.

---

[1] Miller, W. G., The Cobalt-Nickel Arsenides and Silver Deposits of Timiskaming, Ontario, fourth edition, Ont. Bur. Min., Vol. XIX, 1913, Pt. II, p. 48.

The veins of most of the producing mines lay below the diabase sill before it was eroded, e.g., those on the Coniagas, Nipissing, Hudson Bay, Trethewey, Buffalo, Mining Corporation, Crown Reserve, Drummond, Lawson at Kerr lake, LaRose and McKinley-Darragh at Cobalt lake. The King Edward, Silver Cliff, and some of the O'Brien veins lie within the sill. In the outlying camps good examples of veins occurring in the sill are the Wettlaufer of South Lorrain, and Miller-Lake O'Brien of Gowganda.

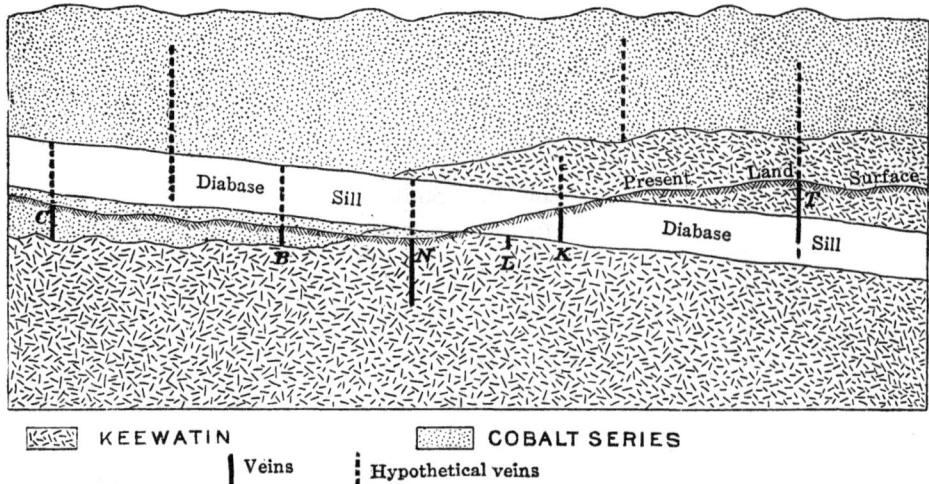

Generalized vertical section through the productive part of the Cobalt area.

The section shows the relations of the Nipissing diabase sill to the Keewatin and the Cobalt series, and to the veins. The eroded surface is restored in the section. The sill is less regular than the illustration shows it to be.

B and C represent a large number of veins that are in the fragmental rocks, Cobalt series, in the lower or foot-wall of the eroded sill. N represents a type of vein, such as No. 26 on the Nipissing, in the Keewatin below the eroded sill, and L a type such as one under Peterson lake, in the Keewatin foot-wall, but not extending upward into the sill; K, a vein in the sill itself, such as No. 3 on the Kerr Lake property; T, a vein such as that on the Timiskaming or Beaver properties, in the Keewatin hanging wall and extending downward into the sill.

At Diabase mountain the top of the hill is diabase, while the rocks below the diabase are composed of slates and conglomerates lying on Keewatin greenstones, so that certain veins, as on the Penn-Canadian and Bailey, started in the sill and continued downward into the underlying rocks.

At the Timiskaming shaft the upper contact between the Keewatin and the diabase is approximately 575 feet from the surface. Along this contact, both above and below, the Timiskaming and Beaver mines have recovered their richest ores. In order to ascertain the thickness of this diabase sill it was diamond-drilled, and the lower contact between the diabase and the Keewatin formations was found at an approximate depth of 1,670 feet from the surface, showing the sill to have a thickness of about 1,100 feet. After diamond drilling was finished, a shaft was sunk through the sill. Exploration work from this shaft, conducted along the bottom of the sill and in the rocks immediately below, has failed to disclose economic ore bodies up to the autumn of 1918.

The following paragraphs, regarding the origin of the ores at Cobalt, are copies from W. G. Miller's report.[1]

The material in the veins at Cobalt has, in all likelihood, been deposited from highly heated and impure waters which circulated through the cracks and fissures of the crust and probably were associated with—followed—the Nipissing diabase eruption.

The waters are said to be associated or connected with the diabase eruption in the sense that they probably represented the end product of the eruption. In many volcanic regions, hot springs are present long after the rocks have solidified. In the Cobalt area the fissures and joints now occupied by the ores were probably produced by the gradual shrinkage in cooling of the diabase, the ores being deposited by the waters which represented the last stage of vulcanicity.

It is rather difficult to predicate the original source of the metals—silver, cobalt, nickel, arsenic, and others—now found in these veins. They may have come up from a considerable depth with the waters, or they may have been leached out of what are now the folded and disturbed greenstones and other rocks of the Keewatin. Analyses of various rocks of the area have not given a clue to the origin of the ores. However, the widespread occurrence of cobalt veins in the diabase, or in close association with it, shown by discoveries throughout a region three thousand square miles or more in extent, appears to be pretty conclusive proof that the diabase and the ores came from one and the same magma.

As the ore bodies in the vicinity of Cobalt station, and elsewhere in Ontario, may be said to be unique among those known in North America, we have no chance of instituting comparisons on this continent. Some European veins, however, such as those of Annaberg, Joachimsthal and other localities [2] show a similar association of minerals.

These European ores are considered by most authors to be genetically connected with intrusions of granite. At Joachimsthal the veins are said to be cut across by basic dikes, and there is evidence to the effect that at the time of the eruption of the dikes the vein formation had not yet been completed. Since especially nickel and cobalt minerals are characteristically connected with basic rocks, the question arises as to whether the European ores mentioned may not be more closely connected in origin with basic rocks than they are considered to be. There may be deeper seated intrusions of these rocks slightly older than the dikes.

Ores and Minerals

The most important ore in the veins at Cobalt is native silver, associated with which is usually some dyscrasite, argentite, pyrargyrite and other compounds of silver, smaltite, niccolite, and related minerals. Many of the minerals occur mixed in the ores and for this reason some of them have not been clearly identified. Another feature of the minerals, which renders their identification difficult, is the fact that most of them occur in the massive form. Crystals when present are small, being frequently almost microscopic in size. The following minerals have been identified and can be conveniently classed under the headings:

I. Native Elements.—Native silver, native bismuth, graphite.

II. Arsenides.—Niccolite, $NiAs$; chloanthite, $NiAs_2$; smaltite, $CoAs_2$; and löllingite, $FeAs_2$.[3]

III. Arsenates.—Erythrite, or cobalt bloom $Co_3As_2O_8.8H_2O$; annabergite, or nickel bloom $Ni_2As_2O_8.8H_2O$; scorodite, $FeAsO_4.2H_2O$.

IV. Sulphides.—Argentite, Ag_2S; millerite, NiS; argyropyrite?; stromeyerite? $(Ag,Cu)_2S$; bornite, Cu_5FeS_4; chalcopyrite, $CuFeS_2$; sphalerite, ZnS; galena, PbS; pyrite, FeS_2.

V. Sulpharsenides.—Mispickel, $FeAsS$; cobaltite, $CoAsS$.

VI. Sulpharsenites.—Proustite, Ag_3AsS_3; xanthoconite? Ag_3AsS_3.

---

[1] Miller, W. G., Ont. Bur. Min., Vol. XIX, 1913, Pt. II, p. 8.

[2] See description, page 6.

[3] Ellsworth, A Study of Certain Minerals from Cobalt, Ontario. Ont. Bur. Min., Vol. XXV, 1916, Pt. 1, p. 223.

3 B.M. (iii)

VII. Antimonides.—Dyscrasite, Ag_6Sb; breithauptite, NiSb.

VIII. Sulphantimonites.—Pyrargyrite, Ag_3SbS_3; stephanite, Ag_5SbS_4; polybasite? Ag_9SbS_6; tetrahedrite, $Cu_8Sb_2S_7$; freibergite? (silver bearing tetrahedrite).

IX. Sulphobismuthites.—Matildite, $AgBiS_2$, emplectite, $CuBiS_2$.

X. Mercury.—Amalgam (?).

XI. Phosphate.—Apatite.

XII. Oxides.—Asbolite; heubachite?; heterogenite?; arsenolite, As_2O_3; roselite? $(Ca,Co,Mg)_3As_2O_8.2H_2O$.

XIII. Veinstones.—Calcite, dolomite, aragonite, quartz, barite, fluorite.[1]

The above table contains a few minerals that have been found in only one or two veins and cannot be considered characteristic. Millerite, for instance, is of rare occurrence, and emplectite has been found only in the Floyd mine, near Sharp lake, in the western part of the Cobalt area. Bornite, chalcopyrite, zinc blende, galena, and pyrite are not characteristic of most of the ore, these minerals occurring more frequently in the wall rock or in non-silver bearing ore of the Keewatin. Apatite in recognizable crystals has been found in the ore of only one mine. Mercury appears to occur in the ore of all the mines that contain high values in silver, but whether it occurs only as amalgam or in other forms has not been determined.

A question-mark has been placed after the names of several minerals in the table which have been reported to occur in the veins, but whose identification has not been made complete by chemical analyses or crystallographic measurements. Gold in small quantity has been found in a number of veins, especially in those in which cobaltite or mispickel are characteristic minerals.

Certain shipments from the Timiskaming mine contained copper in economic quantities.[2]

While we have both native silver and arsenides in abundance, the compounds of arsenic and silver occur only in small quantities. Antimony, which is not abundant, is found in some compounds where we would expect to find arsenic, since the latter is so much more common.

One would also expect to find more compounds of bismuth, since this metal occurs in the free state in considerable quantities in some of the deposits. It might also be expected that native arsenic would occur, but so far it has not been found.

Nearly all the chemical groups of minerals found in the celebrated Joachimsthal deposits of Bohemia are present in the Timiskaming ores. The most important exception is uraninite or pitchblende, which came into prominence a few years ago as the chief source of the element radium.

Order of Deposition of Minerals

The following table shows, in descending order from the youngest to the oldest, the general succession in order of deposition of the principal minerals of the Cobalt area proper. There appear to be, however, minor exceptions to this order.

[1] Barite and fluorite have not been found in the veins at Cobalt proper, but they occur with silver-cobalt ores in one or two veins near Elk lake, and in Langmuir township in the southeast part of the Porcupine area. Small veins of barite have also been found in the Nipissing-diabase in Leonard and Lawson townships, in the Gowganda silver area.

[2] Miller, W. G., Ont. Bur. Min., Vol. XIX, 1913, Pt. II, p. 10.

III. Decomposition products, e.g. erythrite or cobalt bloom, annabergite or nickel bloom, and asbolite.

II. Rich silver ores and calcite.

I. Smaltite, niccolite, and dolomite or pink spar.

After the minerals of group I were deposited the veins were subjected to a slight movement. In the cracks thus formed the minerals of group II were deposited. A few veins that escaped the disturbance do not contain silver in economic quantity.

This order of deposition appears to be the same as that of the minerals in the Annaberg deposits of Germany and those of Joachimsthal, Austria.[1]

Messrs. Campbell and Knight[2] subjected specimens of the cobalt-silver ores of Cobalt to examination, using methods employed in metallography. While their results confirm, in a general way, Miller's observations on hand specimens, and on blocks of ore, they have worked out the order of deposition of the minerals in greater detail. They state that, although all of the structures met with in this examination cannot be satisfactorily explained, they point to the following order of deposition for the principal constituents. First came the smaltite, closely followed by the niccolite; other minerals in small amount came down at this time. Then, after a period of slight movement in which the first minerals were more or less fractured, calcite was deposited as a ground-mass. Later came argentite, which was followed by native silver and native bismuth. Lastly came the surface decomposition products, erythrite and annabergite.

Arranged in order, the succession is, then, as follows:

Smaltite, niccolite, period of movement and fracturing, calcite, argentite, native silver, native bismuth, period of decomposition, and finally erythrite and annabergite.

At Annaberg, bismuth ore is thought to have been deposited with the cobalt-nickel minerals and not with the rich silver ore. Moreover, at the time Messrs. Campbell and Knight made their examination of the ores from Cobalt it was not known that two carbonates occur in the gangue, viz., calcite (white) and dolomite (pink). The latter has been found to belong to an older generation than the former.

Any statement as to the form in which the native silver came in solution into the veins must be merely hypothetical. Silver carbonate, Ag_2CO_3, like calcium carbonate, $CaCO_3$, is soluble in excess of carbon dioxide, CO_2. Hence when the calcite, $CaCO_3$, of the cobalt-silver veins was being carried in solution, it does not seem improbable that silver carbonate may have been in solution at or about the same time.

Palmer and Bastin[3] discuss metallic minerals as precipitants of silver and gold, and their experiments show that certain sulphides and arsenides of copper and

[1] Beck, The Nature of Ore Deposits, Weed's translation, pp. 285-289.

[2] Campbell and Knight, Microscopic Examination of the Cobalt-Nickel Arsenides and Silver Deposits of Timiskaming. Economic Geology, Vol. I, 1906, pp. 767-776. The Para-genesis of the Cobalt-Nickel Arsenides and Silver Deposits of Timiskaming. Eng. and Min. Jour., Vol. LXXXI, 1906, p. 1089.

[3] Palmer and Bastin, Metallic Minerals as Precipitants of Silver and Gold. Economic Geology, Vol. VIII, 1913, p. 140.

nickel, e.g. chalcocite and niccolite, precipitate metallic silver very efficiently from dilute aqueous solutions of silver sulphate. However, the more common sulphides, such as pyrite, galena, and sphalerite were relatively inactive as precipitants of silver from aqueous sulphate solutions.

Argentite, proustite, and native silver in hair-like form, appear to be of secondary origin. These minerals are found in vugs in the lower workings of the mines where the ore has become leaner, or below the productive zone in the veins.

The silver-bearing solutions working downward beneath the sill, in the fractured rocks, lost their silver content by precipitation on coming in contact with the cobalt-nickel minerals before a great depth was reached. Hence it is not surprising to find that rich silver ore does not extend to as great a depth beneath the sill as do the cobalt-nickel ores. Practically all the samples of native silver, excepting those that show a crystalline form or occur in veinlets, contain mercury.

Cobalt minerals are also found in areas lying at some distance from the town of Cobalt. The most important deposits occur in South Lorrain, Casey township, and Gowganda. The Lake Superior silver deposits also contain small amounts of cobalt.

Other minor occurrences of nickel-cobalt ores in Canada are given in the "Annual Report of the Geological Survey of Canada," vol. XIV, 1901, pt. H, to 1917.

The following table shows the production of the Cobalt district from 1904 to 1917.

Total Production of Cobalt Mines 1904-1917[1]

Year	Nickel		Cobalt		Arsenic		Silver		Total value
	tons	value	tons	value	tons	value	ounces	value	
		$		$		$		$	$
1904..	14	3,467	16	19,960	72	903	206,875	111,887	136,217
1905..	75	10,000	118	100,000	549	2,693	2,451,356	1,360,503	1,473,196
1906..	160	....	321	80,704	1,440	15,858	5,401,766	3,657,551	3,764,113
1907..	370	1,174	739	104,426	2,958	40,104	10,023,311	6,155,391	6,301,095
1908..	612	....	1,224	111,118	3,672	40,373	19,437,875	9,133,378	9,284,869
1909..	766	....	1,533	94,965	4,294	61,039	25,897,825	12,461,576	12,617,580
1910..	504	....	1,098	54,699	4,897	70,709	30,645,181	15,478,047	15,603,455
1911..	392	....	852	170,890	3,806	74,609	31,507,791	15,953,847	16,199,346
1912..	429	14,220	934	314,381	4,166	80,546	30,243,859	17,408,935	17,818,082
1913..	377	13,326	821	420,386	3,663	64,146	29,681,975	16,553,981	17,051,839
1914..	(a) 90	28,978	(a) 351	590,406	2,030	116,624	25,162,841	12,765,461	13,501,469
1915..	(b) 35	28,353	(b) 206	383,261	2,490	148,379	24,746,534	12,135,816	12,695,809
1916..	(b) 79	59,380	(b) 400	805,014	2,160	200,103	19,915,090	12,643,175	13,707,672
1917..	(b)155	125,071	(b) 337	1,138,190	2,592	608,483	19,401,893	16,121,013	18,028,597
Total.	4,058	283,969	8,950	4,388,400	38,789	1,524,569	274,724,172	151,950,561	158,176,339

[1] Ont. Bur. Min., Vol. XXVII, 1918, p. 16.

(a) Metallic contents of nickel and cobalt oxides respectively.

(b) Metals and metallic contents of all nickel and cobalt compounds.

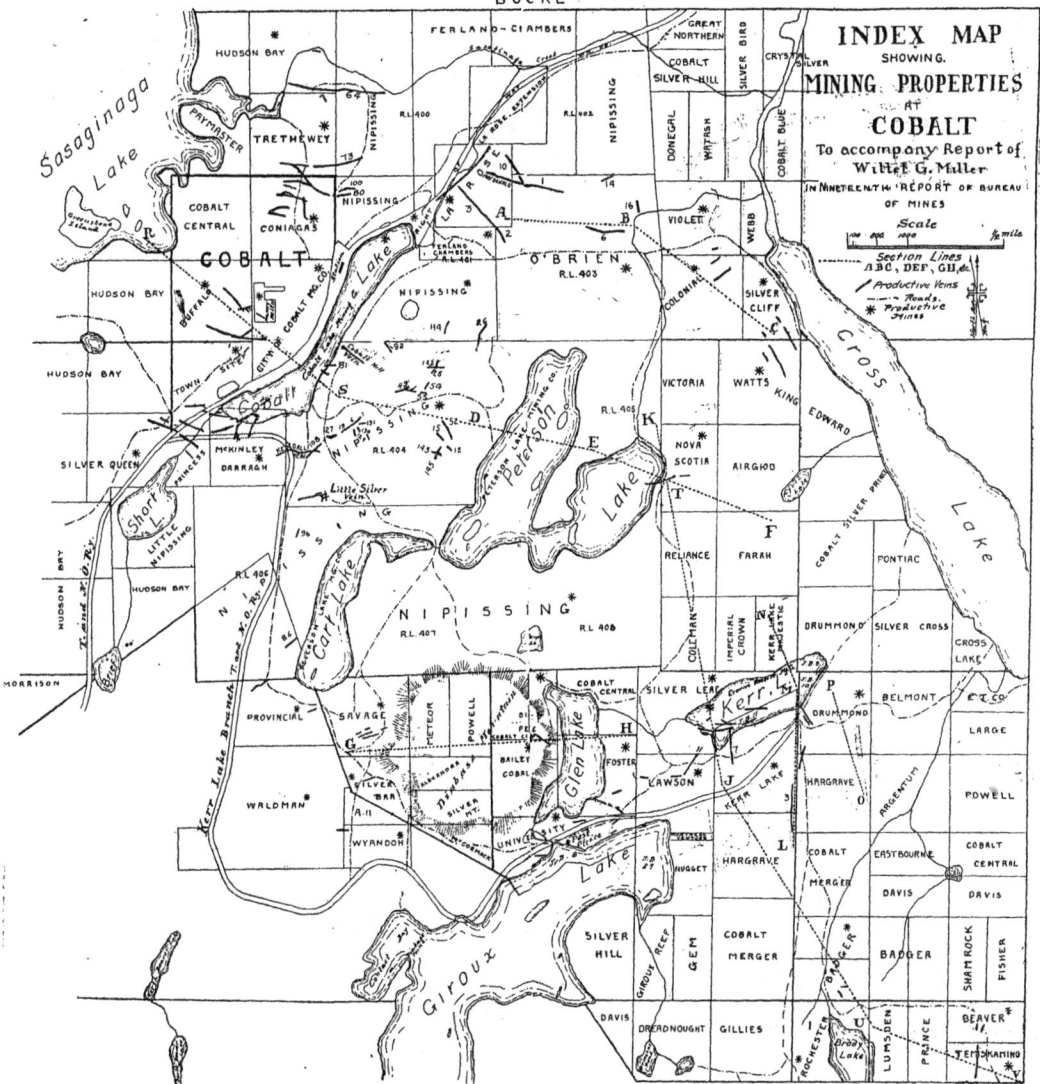

INDEX MAP
SHOWING.
MINING PROPERTIES
AT
COBALT
To accompany Report of
Willet G. Miller
IN NINETEENTH REPORT OF BUREAU
OF MINES

In the following table a record of silver shipments since 1904 is given.

Silver Production, Cobalt Mines, 1904 to 1917[1]

Year	No. of producing mines	Shipments and Silver Contents								
		Ore		Av. per ton	Concentrates		Av. per ton	Bullion	Total	
		tons	ounces	ounces	tons	ounces	ounces	ounces	ounces	value
										$
1904	4	158	206,875	1,309	......	....	......	....	206,875	111,887
1905	16	2,144	2,451,356	1,143	......	....	......	....	2,451,356	1,360,503
1906	17	5,335	5,401,766	1,013	......	....	......	....	5,401,766	3,667,551
1907	28	14,788	10,023,311	677	......	....	......	....	10,023,311	6,155,391
1908	30	24,487	18,022,480	736	1,137	1,415,395	1,244	....	19,437,875	9,133,378
1909	31	27,729	22,436,355	809	2,948	3,461,470	1,174	....	25,897,825	12,461,576
1910	41	27,437	22,581,714	821	6,845	7,082,834	1,030	980,633	30,645,181	15,478,047
1911	34	17,278	20,318,626	1,176	9,375	8,056,189	858	3,132,976	31,507,791	15,953,847
1912	30	10,719	15,395,504	1,436	11,214	9,768,228	871	5,080,127	30,243,859	17,408,935
1913	35	9,861	13,668,079	1,386	11,016	8,489,321	770	7,524,575	29,681,975	16,553,981
1914	32	4,302	6,504,753	1,511	12,152	8,915,958	733	9,742,130	25,162,841	12,765,461
1915	24	2,865	6,758,286	2,359	11,996	10,001,548	834	7,986,700	24,746,534	12,135,816
1916	28	2,177	4,672,500	2,146	8,561	7,598,011	887	7,644,579	19,915,090	12,643,175
1917	28	2,288	3,271,353	1,429	13,720	6,445,243	469	8,053,318	19,401,893	16,121,013
To'l.	....	151,568	151,712,958	1,001	88,964	71,234,197	801	50,145,038	274,724,172	151,950,561

As the camp has developed, the average grade of ore shipped has gradually lowered in value. The introduction of concentration plants in 1908 has tended to keep the shipments up to a high standard, but there is a growing tendency to treat the ore at the mines and recover the silver as bullion for shipment. The average concentration ratio of the different mills during 1914 was 47-1. Further information on the treatment of the ores at Cobalt will be found under the heading " Development of the Metallurgy of the Silver-Cobalt Ores of Ontario."

In the purchasing of the cobalt ores payment is made for the silver and in some cases for the cobalt, the amount paid for the silver varying with the grade of the ore. The different schedules that have been adopted are given in the descriptions of the Coniagas Reduction Co. and the Deloro Mining and Reduction Co. under the " Metallurgy of Cobalt."

In 1905 the price offered for cobalt in ores containing about 6 per cent. cobalt, fell from 65 to 35 cents a pound and at the same time the allowance which had been made previously for the nickel and arsenic, viz., 12 and 0.5 cents a pound respectively was cancelled.

Between 1905 and 1909, ten cents per pound was allowed for the cobalt in the ores if they contained more than 6 per cent., except where the nickel was greater than the cobalt.

Between 1909 and 1914 very little was realized for the cobalt except in the case of high grade ores.

Since 1914, some of the companies have been paying for cobalt, but in some cases not for silver in the same ore. The amount paid for cobalt varies with the

---

[1] Ont. Bur. Min., Vol. XXVII, 1918, p. 16.

grade of the ore and is about as follows: five cents a pound for the cobalt in ores between 6 and 8 per cent., ten cents a pound in ores between 8 and 10 per cent., and fifteen cents a pound in ores over 10 per cent. cobalt.

Most of the cobalt ores that are purchased for the recovery of the cobalt are treated by Canadian smelters. However, a quantity of ore is imported by smelters in the United States, the chief importer being the American Smelting and Refining Company. The Pennsylvania Smelting Co., Carnegie, Pa.; the Balbach Smelting and Refining Co., Newark, N.J.; and the United States Metals Refining Co., Chrome, N.J., also import small quantities of cobalt ores.

Shipments of cobalt-nickel residues from the Nipissing high-grade mill containing 9 per cent. cobalt and 4.5 per cent. nickel have been made by the Nipissing Mining Co. to H. Wiggin and Co., Birmingham, England.

A few shipments containing 4,500 ounces of silver per ton were made previous to 1913 to the Government smelter, Saxony, Germany.

United States smelters imported during 1915, 7,310 tons of ore from the Cobalt district containing 3,580,843 fine ounces of silver, as against 7,206 tons containing 3,966,301 fine ounces in 1914.

In 1916 shipments of ore and concentrates from Cobalt to refineries in the United States comprised 364 tons of ore carrying 408,014 ounces, and 3,700.35 tons of concentrates carrying 1,629,841 ounces—a total of 2,037,855 ounces of silver. In 1917 to refineries in the United States there were consignments from Cobalt amounting to 6,307 tons, from which 2,914,267 fine ounces of silver were recovered. These shipments were on the whole of considerably lower grade than those to the home refineries, averaging only 462 ounces of silver to the ton, as against 810 ounces. Much the larger quantity treated by U. S. plants was at the works of the American Smelting & Refining Company, Denver, Col., and Perth Amboy, N.J. Of the total quantity of silver contained in the product of the Cobalt mines in 1917, namely 19,401,893 ounces, 14,504,681 ounces were refined at the mines in Cobalt or in Ontario works, being about 75 per cent. of the whole.

Additional References

Occurrence and Utilization of Cobalt Ores, Bulletin Imperial Institute, London, Vol. XIV, 1916, pp. 417-437.

Wilson, M. E., Origin of Cobalt Series. Journal of Geology, Vol. XXI, 1913, pp. 121-141.

Power, F. Danvers, The Mineral Resources of New Caledonia, Institution Mining and Metallurgy, Trans., Vol. VIII, 1899-1900, pp. 426-472. This article contains an extensive bibliography.

CHAPTER II

THE METALLURGY OF COBALT

Very little is known about the details of the metallurgy of cobalt in comparison with our knowledge of the other metals, except by those directly connected with the industry. It is not a new subject, since the treatment of cobalt ores was practised for several hundred years in Europe, where the output of the world's supply of cobalt was controlled until the discovery of the Canadian cobalt deposits in 1903. New South Wales, Norway, New Caledonia, Germany, Chile, and Hungary were the chief producers of cobalt ores, while the largest refineries were located in Germany and England. Since 1902 there has been very little cobalt ore mined outside of Canada, except in the United States during 1903 and 1908, when there was a production of 60 and 100 tons respectively of cobalt oxide from the ores of Missouri. Until 1913, the world's annual production of cobalt oxide amounted to approximately 250 tons, but within recent years the production has increased until in 1916 it amounted to 400 tons. Within the last few years the quantity of cobalt metal produced has increased from practically nothing in 1913, to 165 tons in 1916, and 158 tons in 1917.

The price of cobalt oxide (70 per cent. cobalt) fluctuated little previous to 1907, the oxide selling at prices varying from $1.60 to $2.00 a pound. In 1907 the price rose to $2.50, but in 1908 it dropped to $1.40. Since 1908 the price has gradually declined, the average for 1915 being 90 cents a pound. Owing to the increased present demand the price has risen to $1.50 (1917). The value of metallic cobalt is given (1917) as $2.00 to $2.25 a pound.

In reviewing the metallurgy of cobalt, two noticeable changes are evident; first, previous to the discovery of the large cobalt deposits in Canada, practically all compounds of cobalt were produced in Europe; and second, in the European refineries ores were treated for the cobalt content alone, while from the ores of Canada, metallic silver, cobalt, nickel, and arsenic oxide are recovered. The associated metals are often a source of revenue for the smelters.

Since most of the cobalt compounds produced in Europe were used in the ceramic industries, and as the requirements of these industries at the time were not such as to demand a high-grade cobalt oxide, it is reasonable to conclude that the processes used in Europe did not produce a high-grade cobalt oxide. However, the demand of the ceramic industries at the present time is for a high-grade oxide, and this is supplied by the Canadian smelters at practically one-half the price that the low and medium cobalt compounds or smalts were sold at in Europe ten years ago.

The elements arsenic, sulphur, copper, iron, and nickel, which are usually associated with cobalt ores, are common to the ores of Europe and Canada, while those from New Caledonia, though free from sulphur and arsenic, contain a large percentage of manganese. However, arsenic and sulphur cannot be altogether considered as impurities in cobalt ores, since the presence of either element enables the ores to be reduced in blast-furnaces to produce a speiss or matte.

The metal cobalt or the oxide has never been recovered in a pure form from ores by dry methods alone, because of its association with metals possessing very similar properties; hence we find chemical or wet methods employed to separate the associated elements. In any preliminary treatment or smelting of cobalt ores the behaviour of cobalt, nickel, and iron toward arsenic, sulphur, and oxygen is important. Of the three metals nickel, cobalt, and iron, nickel has the greatest affinity for arsenic, then cobalt, and lastly iron; while in the case of oxygen their affinities are reversed. As regards the behaviour of these metals towards sulphur, there is little difference, but nickel and cobalt seem to combine preferably with sulphur. Also, because the affinities of the three metals lie very close together, it is not possible to eliminate iron as oxide or silicate from a mixture of iron, nickel and cobalt by having just sufficient arsenic present to form arsenides of nickel and cobalt, nor is it possible to remove iron from the other two in a speiss or matte by regulating the extent of oxidation. In both cases some iron, nickel, and cobalt will be found together.

Cobalt ores are commonly smelted in blast furnaces to remove gangue minerals and other impurities, e.g., iron, sulphur, and arsenic. In the blast furnace smelting, a speiss[1] or a matte[2] and a slag are formed. In blast-furnace smelting of cobalt ores a certain amount of iron is always allowed to enter the speiss or matte, because when iron is present very little cobalt will be found in the slag, while a certain amount of iron is necessary to assist in the subsequent precipitation of arsenic.

The following classification illustrates the metallurgical treatment of cobalt ores. Although the production of smalt[3] directly from ore is not practised at the present time, this method is given in the classification since it was used formerly in Europe.

1. The Extraction of Cobalt Oxide.
    A. Decomposition of arsenical and sulphide ores;
        1. By smelting in blast-furnaces producing,
            (a) A speiss containing chiefly cobalt and nickel arsenides.
            (b) A matte containing chiefly cobalt and nickel sulphides.
        2. By other processes.
            (a) Wet processes.
            (b) Dry processes.
    B. Decomposition of oxidized ores;
        1. Wet processes.
        2. Dry processes.
    C. Decomposition of Silicates.
2. The Production of Smalt.

I.—The Extraction of Cobalt Oxide

A. 1. Decomposition of Arsenical and Sulphide Ores in Blast Furnaces

All the cobalt ores smelted in Canada are arsenical ores containing silver, while those treated in Missouri are sulphide ores practically free from silver.

---

[1] Speiss is a metallurgical product in which the metals are present as arsenides.
[2] Matte is a metallurgical product in which the metals are present as sulphides.
[3] Smalt is a silicate of cobalt, used in the pottery industries.

From the arsenical silver ores, metallic silver and argentiferous speiss containing cobalt, nickel, iron and copper as arsenides are produced, while from the sulphide ores a matte containing the metals as sulphides is formed. If more than sufficient arsenic or sulphur is present than is necessary to combine with the cobalt, nickel, copper and part of the iron, any excess is removed by a previous roasting, or a number of different ores may be mixed to get the proper quantity of arsenic for the blast furnace charge. The usual blast furnace charge contains approximately 16 per cent. arsenic.

In the blast-furnace smelting of the silver cobalt ores of Canada, the products of smelting are metallic silver, approximately 850 fine, which represents about 85 per cent. extraction of the silver content of the ore; an argentiferous speiss containing arsenides of cobalt, nickel, copper, and part of the iron; a slag containing the lime, magnesia and part of the iron as silicates; and flue dust which contains fine particles of ore, crude arsenious oxide, and coke dust. In smelting sulphide ores, free from silver, matte, slag, and flue dust are produced. The speiss or matte then undergoes further treatment for the recovery of the cobalt, nickel, and silver.

The process of treating speiss or matte consists of grinding and roasting, followed by treatment with sulphuric acid and chemicals to convert the cobalt and nickel into soluble compounds, leaving most of the impurities in an insoluble state. However, small quantities of iron, nickel, copper, arsenic, and sulphur dissolve with the cobalt, and these must be removed, since they are objectionable in cobalt oxide to be used in the ceramic industries. The maximum limits of the above impurities in high-grade cobalt oxide are, approximately, iron 0.5 per cent., nickel 1.0 per cent., copper, arsenic, and sulphur 0.1 per cent each.

Treatment of ground unroasted speiss with acids and chemicals is also practised.

The smelting of arsenical ores of cobalt is conducted in blast-furnaces at present in Canada, chiefly at the smelters of the Deloro Smelting and Refining Company and the Coniagas Reduction Company. The blast-furnace smelting of sulphide ores is used by the Missouri Cobalt Company, Fredericktown, Missouri, which was formerly the North America Lead Company. Sulphide ores were also treated at one time at the Scopello works, Piedmont, Italy; at the Isabella works, in Silesia; at Schneeberg in Saxony; and at the Christofle works, at St. Denis, France.

A. 2 (a). Decomposition of Arsenical and Sulphide Ores by Wet Processes

No methods have ever been successfully practised to treat arsenical and sulphide cobalt ores directly by wet methods alone, the reason being that it is more profitable to concentrate this class of ores by producing a speiss or matte. A few attempts have been made to treat arsenical ores without a preliminary smelting in blast-furnaces, by first roasting the ore, which operation was followed by treatment with acids. Owing to the difficulties in operating such a process, and also to the high consumption of acids and chemicals, only small quantities of ore containing small amounts of soluble gangue minerals could be treated. The reagents tried were hydrochloric, sulphuric, or nitric acids, with or without the addition of chemicals. Solutions of ferrous chloride were also frequently tried.

4 B.M. (iii)

A. 2 (b).   Decomposition of Arsenical and Sulphide Ores by Dry Processes

An attempt was made in Germany to treat low-grade cobalt ores by roasting with the addition of salt and iron pyrites, the cobalt being converted into soluble cobalt chloride, while practically all the iron remained as insoluble oxide.   There is no record of such a process ever having been operated on a commercial scale.

B. (1).   Decomposition of Oxidized Ores by Wet Processes

The treatment of the ores from New Caledonia comes under this heading. Decomposition of the ore was effected by treatment with a hot solution of ferrous sulphate.   For a detailed description of this method see the Herrenschmidt processes.

B. (2).   Decomposition of Oxidized Ores by Dry Processes

An attempt was made during 1893 and 1894 to concentrate the oxidized ores of New Caledonia by producing a matte.   As these ores did not contain sufficient sulphur to form a matte, this latter element must have been added, probably as pyrites.

C. Decomposition of Silicates

Silicates of cobalt cannot be smelted with arsenical or sulphide ores to form a speiss or matte, because cobalt silicate is not decomposed by iron arsenide or iron sulphide to form cobalt arsenide or sulphide.   Nickel silicate, however, does react with iron pyrites to give nickel sulphide and iron silicate.   It is possible, however, under strongly reducing conditions, to reduce some cobalt from the silicate.

The possibility of treating silicates by wet methods will depend altogether on whether or not the cobalt and gangue minerals are decomposable by acids.   However, there are no known occurrences of cobalt silicates in nature.

As mentioned above, cobalt ores usually contain arsenic, sulphur, copper, iron, nickel, and manganese.   The larger part of the arsenic, sulphur, and iron is removed in the preliminary blast-furnace treatment, but the chief difficulty in producing fairly pure cobalt oxide lies in the removal of the remaining small quantities of these elements.   The removal of these impurities will be discussed in the order in which they are usually removed from cobalt-nickel solutions.

Removal of Arsenic from Cobalt-Nickel Ores and Solutions

Arsenic is a common impurity of cobalt ores, and is very objectionable, especially in cobalt metal which is added to other metals to make alloys.   When dry or wet methods are employed to convert cobalt and nickel into soluble form, considerable arsenic dissolves with the former metals, and any arsenic in solution must be removed before proceeding with the precipitation of the cobalt and nickel.

The method of removing arsenic depends chiefly on the quantity and form in which it is present.   In case of large quantities of arsenic, roasting to convert a greater part of it into volatile arsenious oxide is practised, but it is difficult to roast an ore containing more than 20 per cent. of arsenic to below 7 per cent., because of the formation of arsenates of iron, cobalt, and lime.   The addition of carbon reduces any arsenates to arsenites, from which compounds volatile arsenious oxide is evolved on heating.   This method, however, is too costly for any extensive use.

In case of roasted products or ores containing small quantities of arsenic in the form of arsenates, chemical methods must be used to remove this impurity.

When cobalt ores are treated in blast furnaces, considerable iron is allowed to enter the speiss or matte, so that the loss of cobalt and nickel in the slag will be low. A part of this iron is dissolved with the cobalt and nickel. Any dissolved iron assists in the removal of arsenic, for when a solution containing nickel, cobalt, iron, and arsenic is neutralized with ground calcium carbonate, the iron and arsenic combine to form ferric arsenate, $FeAsO_4$, a light-brown, flocculent to granular precipitate. Practically all the arsenic can be removed in this way, and if there is not sufficient iron in the solution to form ferric arsenate with all the arsenic, iron in some soluble form is added.

Arsenic may also be removed from speiss by heating the roasted product with sodium carbonate and nitre to form soluble sodium arsenate, the metals forming oxides. This method was practised for some time by the Canadian Copper Company, Copper Cliff, Ontario, in the treatment of the silver-cobalt ores. The leached speiss containing up to 3 per cent. arsenic was shipped to New Jersey to be refined.

Removal of Iron

Any iron not removed as ferric arsenate is precipitated by careful additions of calcium carbonate, the iron precipitating as ferric hydrate. To completely precipitate the iron it is essential that the iron be oxidized.

Removal of Copper

Copper, when present in cobalt-nickel solution in quantities over 0.5 per cent., presents considerable difficulty in its complete removal. Small amounts of copper will be precipitated completely from cobalt-nickel solutions along with the iron by the addition of calcium carbonate. However, when the ratio of the copper to the cobalt and nickel is greater than 1 to 6, it is advisable to remove the copper, either by iron plates or electrolytic methods.

Removal of Manganese

There is practically no manganese in the cobalt ores found at Cobalt, Canada, but practically all the cobalt ores from New Caledonia, previously in large use, carried a high percentage of this metal. In the manipulation of the New Caledonia ores the solutions containing cobalt, nickel, and manganese were treated with sodium sulphide, the cobalt and nickel being precipitated as sulphides, while practically all of the manganese remained in solution.

Separation of Cobalt from Nickel

After the arsenic, iron, copper, and manganese have been removed, the next step is to separate the cobalt and nickel. The commonest method is to precipitate the cobalt first, but it is possible to precipitate it after the nickel.

The separation of cobalt from nickel as practised at the present time is practically the same as it was a number of years ago. When a solution of calcium hypochlorite (bleaching powder) is added to a solution containing cobalt and nickel, the cobalt is precipitated first as a black hydroxide, $Co(OH)_3$. The precipitation may be carried to a colourless solution[1] without precipitating any appreciable quantities of nickel. If it is desired to obtain pure nickel oxide, the first cobalt hydroxide is removed, and the precipitation of the remaining cobalt is continued until the solution is practically free from it, a quantity of black nickel hydroxide, $Ni(OH)_3$, being precipitated at the same time. This intermediate precipitation produces mixed oxides, which must be retreated to produce pure cobalt and nickel oxides. The nickel in solution is precipitated as nickelous hydroxide or hydrated carbonate by the addition of a solution of lime or sodium carbonate.

When cobalt and nickel are present as sulphates, it is customary to precipitate the cobalt by sodium hypochlorite instead of calcium hypochlorite, since the lime of the calcium hypochlorite reacts with the sulphate radicle to form insoluble calcium sulphate, which is difficult to remove. In case calcium sulphate is present in cobalt oxide, it may be removed by a treatment with a hot solution of sodium carbonate, the sodium carbonate and calcium sulphate reacting to form sodium sulphate and calcium carbonate. The sodium sulphate is removed by water and the calcium carbonate by treatment with dilute hydrochloric acid.

Cobalt may also be precipitated after the nickel. In this case soda is added to the boiling solution which precipitates nickel hydroxide and carbonate with a small amount of cobalt carbonate, while cobalt with a small amount of nickel remains dissolved. Cobalt is finally precipitated by additions of more soda or chloride of lime. The writer is not aware of this method being practised commercially.

Methods Used or Proposed to Treat Cobalt Ores for Cobalt Oxide

In the following pages of this section a brief outline is given of the methods practised successfully at the present time and those formerly employed in Europe. This is followed by a summary of all the processes that have been proposed to treat cobalt ores. Very few, if any, of these latter processes could ever be successful on a practical scale, while some are merely laboratory methods. An outline of all the processes is incorporated in this review merely as a reference for anyone undertaking an investigation of the recovery of cobalt from its ores. A more detailed description is given of the processes and smelters of the Coniagas Reduction Company, Limited, and the Deloro Smelting and Refining Company, Limited, since these two smelters are the most successful ones at the present time. The Herrenschmidt process is also considered in detail as a large quantity of New Caledonia ores were formerly treated by this method. All the other processes are grouped together, mention only being made of differences in the proposed treatments that may possibly be of interest.

The methods employed by Canadian companies are first described.

---

[1] A solution containing approximately 1.5 to 2 parts of nickel sulphate to 1 part of cobalt sulphate is practically colourless. If cobalt is present in a larger ratio, the solutions are pink to red in colour, while if the ratio of the cobalt to nickel is less the solutions are green.

Coniagas Reduction Company Limited[1]

The Coniagas Mines Limited, of Cobalt, Ontario, owns practically all of the issued capital stock of the Coniagas Reduction Co., Limited. The head office of the company is at St. Catharines, Ontario, but the smelter is situated at Thorold, six miles west of Niagara Falls. The company's property comprises 160 acres, of which the smelter occupies about four, with a frontage of 1,500 feet on the Welland canal. It is also served by the Grand Trunk, and Niagara, St. Catharines, and Toronto railways.

The construction of the smelter was begun in March, 1907, and actual smelting commenced in May, 1908. It was erected for the treatment of ores from the Coniagas mine, but its capacity is sufficient to smelt a certain tonnage of other silver ores from Cobalt.

The process in use at this smelter is as follows: The ore is crushed, ground in a Krupp ball-mill, and sampled by a Vezin automatic sampler, two separate samples being taken. The ground ore is smelted in a blast-furnace with limestone and iron ore, the products being impure metallic silver, an argentiferous speiss containing cobalt, nickel, and iron as arsenides, also flue dust, and slag. The impure silver is cast into anodes and refined electrolytically. The speiss is treated with chemicals to recover the silver, cobalt, and nickel. Various grades of cobalt oxide containing from 60 to 76 per cent. metallic cobalt, are produced, according to the demand of the market. The cobalt oxide contains less than 1 per cent. nickel and only small proportions of sulphur, lime, and iron. The arsenical fume from the dust-flues and collectors is treated to produce refined white arsenic, which assays over 99 per cent. arsenious oxide.

To operate the plant, from 200 to 300 horse-power is required, which is transmitted from Niagara Falls. The smelter has a monthly capacity of 250 tons of raw ore. The limestone flux is obtained from Port Colborne, 20 miles south, and the iron ore from Michigan.

This company produces refined silver, cobalt oxide and metal, nickel oxide and metal, white arsenic and metallic arsenic.

The output of the smelter since the commencement of operations is given below. The production of cobalt and nickel oxides, as shown in the table, represents the cobalt and nickel content in refined oxides and various products.

Year	Ore Treated	Silver—Fine	Cobalt Oxide	Nickel Oxide	White Arsenic
	tons	ounces.	tons	tons	tons
1908....................	266.80	360,683	5.5	1.5	13.5
1909....................	1,116.90	1,659,604	0.9	....	100.0
1910....................	2,017.25	3,485,243	53.8	13.2	557.7
1911....................	2,821.50	5,770,271	60.5	17.3	766.1
1912....................	2,288.77	4,824,632	129.0	50.7	636.7
1913....................	2,509.80	4,977,012	250.6	115.6	319.4
1914....................	1,968.78	3,865,546	171.9	124.9	399.2
1915....................	2,541.00	3,445,661	59.0	99.8	472.8
1916....................	2,718.86	4,428,913	190.4	67.6	420.8
1917....................	2,633.25	2,954,665	49.6	38.9	555.3

[1] Cole, Arthur A., Report of the Timiskaming and Northern Ontario Railway Commission, Toronto, Canada, 1912, p. 69.

The smelting schedule of the Coniagas Reduction Company in condensed form is as follows:

Schedule.—Percentages of silver to be paid for on commercial assay of the silver content per ton of 2,000 pounds as follows:

55 per cent. for	50 ounces and proportionate increase in percentage up to	200
73 " "	200 " " " " " "	300
78 " "	300 " " " " " "	500
84 " "	500 " " " " " "	1,000
91.5 " "	1,000 " " " " " "	1,500
92.5 " "	1,500 " " " " " "	2,000
93.5 " "	2,000 " " " " " "	3,000
95 " "	3,000 and over.	

Sampling to be at vendor's expense.

All ore purchased to be subject to a refining charge of 0.75 cent per ounce of silver content.

Payment:—75 per cent. of the amount 30 days after date of weighing and sampling reports; 25 per cent. of amount 90 days after date of said report.

Price of silver to be determined by New York quotation as given by Messrs. Handy and Harman to Western Union Telegraph Company on dates of settlement.

Deloro Smelting and Refining Company Limited[1]

The Deloro Smelting and Refining Company[2] is a close corporation controlled by M. J. O'Brien,[3] owner of the O'Brien mine, Cobalt, and Miller-Lake O'Brien mine, Gowganda.

The smelter is located at Deloro, Hastings county, Ontario, one mile from Marmora station, on the Canadian Northern railway.

The plant was built and operated first as an arsenic refinery by the Canadian Goldfields, but was entirely remodelled in 1907 by the present owners to smelt ores from Cobalt, particularly those of the O'Brien mine. During the year 1908 a separate and extensive plant was added for the production of cobalt and nickel oxides, and this has been in successful operation since May, 1910. When the plant was first erected the products were limited to silver, refined arsenic, and mixed oxides, but it has beeen gradually extended and at present the company produces refined silver, cobalt oxide and metal, nickel oxide and metal, and white arsenic. There is also an equipment for the production of the cobalt-chromium-tungsten alloy known as stellite, used for high-speed cutting-tools.

Treatment of Ores.—Ores and mill products from Cobalt are purchased on a basis of the silver content. Sampling is done carefully under the supervision of a representative of the seller, the process in use being as follows: Each carload of ore is stored in a separate bin, from which it is removed and crushed to 15-mesh in a ball-mill, to which is attached a Snyder sampler. This machine takes about 50 samples a minute, each one representing 10 per cent. of the ground material leaving the mill. The total sample is subdivided until a final sample of about 20 pounds is obtained. The coarse scales of silver which do not

---

[1] This company was formerly known as the Deloro Mining and Reduction Company, Limited.

[2] Cole, Arthur A., Report of the Timiskaming and Northern Ontario Railway Commission, Toronto, Canada, p. 74, 1913.

[3] Now M. J. O'Brien, Limited.

pass through the ball-mill screens are melted and cast into a bar which is weighed, sampled and assayed. The final assay of the ore is calculated from the assays of the ore and scales.

The following are typical assays of ores and mill products received at the smelter up to 1915:

Product	Ag per ton Oz.	Co	Ni	Cu	Fe	As	S	SiO$_2$	CaO	MgO
Ore (hand–picked).......	2,194	7.9	4.3	0.10	5.0	30.2	1.70	4.17	15.0	2.7
Jig product	1,442	10.4	5.8	0.20	6.5	47.2	3.70	4.5	5.2	0.8
Table concentrate.......	1,426	8.2	3.8	0.25	11.6	37.1	8.25	9.5	......	......
Slime "	324	2.1	0.5	......	6.8	10.0	2.98	58.3	2.5	1.92

The ground ore with the required fluxes is mixed in a pug-mill and smelted in a low-pressure blast furnace, the products being metallic silver, an argentiferous speiss, slag, and flue dust.

The silver, which is about 850 parts fine, is charged into an oil-fired refining furnace and the impurities oxidized. Silver 996 parts fine is produced.

The argentiferous speiss is re-crushed, roasted in coal-fired reverberatory furnaces or in an oil-fired Brückner furnace, and the product conveyed to the chloridizing furnaces, where it is heated with salt.

The chloridized speiss is charged into agitating tanks where the silver is extracted by sodium cyanide. Metallic silver is precipitated from the cyanide solution by the addition of aluminum dust. This process was developed by Prof. S. F. Kirkpatrick, Queen's University, Kingston, Canada. The silver obtained is exceptionally high-grade, and the cyanide is to a large extent regenerated. The silver from the refining furnace is mixed with the silver precipitate from the cyanide process and treated with borax and nitre in an oil-fired tilting furnace, after which it is poured into moulds.

The residues from the cyanide treatment are given further treatment for the separation by precipitation of the cobalt and nickel.

Power is supplied to the smelter from Campbellford by the Hydro-Electric Power Commission of Ontario, over a 22-mile transmission line, at $20 per horse-power year. To operate the plant 300 to 400 horse-power is required.

The following table shows the production of the Deloro smelter since 1908:

Year	Ore Treated tons	Silver, Fine ozs.	Cobalt and mixed Oxides tons	Refined Arsenic tons
1908 to 1912....................	11,065	20,339,860	500	3,275
1913............................	2,920	6,350,500	190	893
1914............................	3,612	5,207,000	300	1,038
1915............................	4,634	6,429,794	256	1,634
1916............................	3,553	5,234,620	....	1,627
1917............................	3,086	3,474,613	....	1,809

Schedule.—Payment is made for 98 per cent. of the silver content of the ore determined by commercial assay, on the following terms and conditions:

Treatment charge, $25 a ton of ore.

Refining charge, 0.75 cent per ounce of silver content on ore assaying 3,000 ounces and over per ton; one cent per ounce on ore assaying 2,000 to 3,000 ounces per ton; one and one-half cents per ounce on ore assaying less than 2,000 ounces per ton.

Terms of payment, 75 per cent. of net proceeds 30 days after completion of sampling. All ore is to be delivered in carload lots f.o.b. Marmora station, and to be at the shipper's risk until sampling is undertaken.

Canadian Copper Company

The cobalt plant of the Canadian Copper Company was situated at Copper Cliff, about one-quarter of a mile south of the large copper-nickel smelter of the same company.[1]

The works were designed to smelt and treat ores and concentrates from the Cobalt silver mines, and were in operation from 1905 to 1913. This plant was closed because of the extended treatment of the ores in cyanide plants at the mines. The following is an outline of the method used:

Treatment.—The ore was crushed, ground in a ball-mill to 30-mesh, and from the ground ore, one-tenth was removed by a Snyder sampler. Sampling was completed by coning and quartering. The first quartering divided the sample into two parts, which were treated as two independent samples. The ore was charged with suitable fluxes into a 30 by 72-inch blast furnace, having a capacity of 25 to 30 tons per 24 hours. Limestone from Michigan was used as a basic flux, and low-grade cobalt ore as an acid flux when required.

The products of the blast furnaces were: silver, speiss, slag, and flue dust containing fine ore, coke, and crude arsenic oxide.

The silver and speiss were tapped from the furnace through the lower tap-hole and allowed to settle in a slag pot, the silver going to the bottom.

The silver button, assaying 850 parts fine, and weighing about 50 to 75 pounds, represented an extraction of about 75 per cent. of the silver in the ore. The grade of the silver was raised to 980 parts fine in an oil-fired refining furnace which had a capacity of 30,000 ounces. It was shipped in bars to the Balbach Smelting and Refining Company, Newark, N.J., for further refining. The slag from the refining furnace was returned to the blast furnace.

An average analysis of the speiss produced is as follows: silver 900 oz. per ton, arsenic, 24 to 30 per cent., cobalt 27, nickel 9 to 15, iron 20, copper 2, and sulphur 6.

Previous to 1909 the blast furnace speiss was ground and roasted to remove most of the arsenic. The roasted speiss was mixed with sodium carbonate and nitre and again heated. By this treatment the arsenic was changed to soluble sodium arsenate, the cobalt and nickel to oxides. After removing the soluble arsenate, the residue, termed leached speiss, containing 2 to 3 per cent. arsenic and considerable

---

[1] Bridges, The Metallurgy of Canadian Cobalt Ores. *Con. Min. Jour*, Vol. XXXVII, 1916, pp. 48, 68, 134.

silver, was shipped to New Jersey to be refined. The above method of treatment was changed about 1909 in order to recover more of the silver, the new process being as follows:

The speiss was ground to 30-mesh, mixed with 20 per cent. salt, and roasted in mechanically rabbled Edwards furnaces, fitted with water-cooled rabbles. Each furnace had a capacity of 2,400 pounds per 24 hours. The chloridized speiss was then treated with water to dissolve the soluble cobalt, nickel, and copper salts. The solution was passed through a tank containing scrap iron, which precipitated the copper, after which the cobalt and nickel were precipitated as hydroxides by caustic soda, converted into oxides in an oil-fired furnace, ground in a pebble-mill, and barrelled for shipment. An approximate assay of this material is as follows: silver 15 oz. per ton, arsenic 0.3, cobalt 40, and nickel 3 per cent. The small amount of nickel in the foregoing analysis in proportion to the cobalt is due to the nickel chloride being more readily decomposed than the cobalt chloride. The treatment of the speiss was continued with four covers of hyposulphite of soda solution, the residue finally containing 20 to 30 oz. of silver per ton. The silver was precipitated as sulphide by the addition of a saturated solution of sodium sulphide, filtered in a filter-press, dried, mixed with 100 per cent. sodium nitrate and 10 per cent. sodium carbonate, heated to redness in an oil-fired roasting furnace, and then transferred to tanks where it was leached with hot water. A spongy mass of metallic silver remained, with a small quantity of cobalt and nickel. The spongy mass, which contained from 60 to 65 per cent. of silver, was added to the bath in the silver-refining furnace.

The residues from the first hyposulphite leaching were mixed with quartz and smelted in a blast furnace to remove the iron. The resultant products were slag, speiss, and flue dust.

The slag, which contained 15 ounces silver per ton, 10 per cent. cobalt, and less than 1 per cent. nickel, was smelted with other high-silver slags, and pyrite from Capelton, Quebec.

The speiss from this second smelting had the following approximate composition: silver 300 ounces per ton, arsenic 25 to 30 per cent., cobalt 35, nickel 25, iron 3.5, and copper 2 per cent., also a little sulphur when the arsenic was low.

The second speiss was treated similarly to the first up to the time when the first residue was removed from the cylinders after treatment with hyposulphite. The second speiss residue, which contained 20 per cent. arsenic, was mixed with 20 per cent. sodium nitrate and 10 per cent. sodium carbonate, and roasted in a hand-rabbled reverberatory furnace. This treatment changed the arsenic to sodium arsenate, which was dissolved in hot water, the solution being discarded. The residue, after drying, had the following approximate composition: silver 20 to 30 oz. per ton, arsenic 0.3 to 0.7 per cent., cobalt 35 to 37, nickel 23 to 25, copper 3, and iron 5 per cent.

Payment was received for the silver in the residues, as well as for the cobalt and nickel oxides.

The arsenious oxide from the blast furnace and roasting furnaces was collected in flues and charged into an arsenic refining furnace. The residue, a clinker high in silver, was returned to the blast furnace. The final product was refined white arsenic which contained 99.98 per cent. arsenious oxide (As_2O_3) and 0.3 ounces of silver per ton.

The slag from the blast furnace was rejected except when it assayed more than 10 ounces of silver per ton, in which case it was retreated in the blast furnace.

The 200 to 300 horse-power required was supplied from the company's plant at High Falls, 14 miles from the smelter.

The following table shows the ore treated and the production of the cobalt plant of the Canadian Copper Company from the commencement of operations to their close in 1913.

| Year | Ore Treated Lbs. | Silver, Fine Ozs. | Metallic[1] | | White Arsenic lbs. |
			Cobalt lbs.	Nickel lbs.	
1906...............	1,767,692.5	1,282,692.78	9,021	3,987	....
1907...............	4,560,627.5	3,829,542.82	331,151	138,427	510,622
1908...............	9,857,072.5	8,551,582.07	464,171	268,140	942,827
1909...............	10,651,189.5	8,779,014.55	690,737	463,588	1,242,722
1910...............	9,792,511.0	8,696,624.87	346,483	260,756	843,619
1911...............	6,744,108.0	6,584,102.46	238,684	234,323	680,074
1912...............	3,667,301.0	3,523,207.80	223,163	209,330	476,156
1913...............	186,602.0	47,590.00	15,506	7,161	95,669
Total	47,227,104.0	41,294,357.35	2,318,916	1,585,712	4,791,689

Canada Refining and Smelting Company Limited

The plant of the Canada Refining and Smelting Company Limited,[2] was situated in the southern part of the town of Orillia, Ontario, and adjacent to the Grand Trunk, Canadian Pacific, and Canadian Northern railways.

Construction was started early in September, 1910, and smelting was commenced in February, 1911.

The plant was designed for the treatment of silver ores from Cobalt, and had a capacity of about 13 tons daily. It produced refined silver, white arsenic, and mixed oxides of cobalt and nickel.

This plant has not been operated since early in 1913, but the treatment of the ore was as follows:

The crushing and sampling was done at Cobalt by Campbell and Deyell, samplers and assayers, before shipment to the smelter. The ore was smelted in a 48-inch circular shaft furnace, which produced silver, argentiferous speiss, slag, and flue dust.

The silver recovered from the furnace assayed 900 parts fine and represented an extraction of 80 per cent. It was refined to 996 parts fine in a cupellation furnace of a capacity of 10,000 ounces. The slag from the refining furnace

---

[1] These figures represent the metallic nickel and cobalt contained in the crude oxides in which form they were shipped.

[2] Cole, Arthur A., Report of the Temiskaming and Northern Ontario Railway Commission, Toronto, Canada, p. 68, 1913.

reverted to the blast furnace. Limestone and iron ore were used as fluxes when required, the limestone being obtained from Longford quarry, nine miles distant from the smelter, and the iron ore brought from Midland, Ontario.

The speiss was ground, roasted and re-ground. It was treated with chemicals, when most of the metals except the silver were dissolved. The impure silver-bearing residue was separated from the liquor in filter-presses and recharged into the cupola furnace.

The iron, arsenic, and copper were first precipitated from the liquor, and finally the cobalt and nickel precipitated together as carbonates. The mixed carbonates were heated in a hearth furnace and converted to oxides which, after being ground, were barrelled and shipped. The oxides assayed 40 per cent. of cobalt and 25 per cent. of nickel.

The flue dust was treated to recover pure arsenious oxide.

About 300 horse-power was required by the plant. This was supplied by the town of Orillia, from a hydro-electric installation 18 miles distant from the town, at the rate of $18.40 for 24-hour service per horse-power year. About 80 men were employed.

Metals Chemical, Limited

During 1915 Metals Chemical Limited, Welland, Ontario, erected a plant to treat low-grade Cobalt ores and residues. The ores are smelted in a blast furnace, with a capacity of 30 tons of ore per day. The products of the furnace are silver, speiss and slag. The speiss is roasted to remove the arsenic and the roasted product treated with chemicals to dissolve the cobalt and nickel. The following products are shipped: cobalt oxide, cobalt carbonate and sulphate, nickel oxide and sulphate, refined silver and white arsenic (As_2O_3).

The Standard Smelting and Refining Company Limited, erected during 1914, a small plant at North Bay, Ontario, to treat ores from Cobalt. This company in 1915 moved its works to Orillia, Ontario, and cobalt ores were treated during part of 1916. The company went into liquidation and the plant was taken over by the International Molybdenum Co.

Under the provisions of the " Metal Refining Bounty Act," passed by the Ontario Legislature in 1907, there was paid a bounty of six cents a pound on the metallic contents of cobalt and nickel oxides produced within the Province. The total bounties to be paid in any one year were not to exceed the sum of $30,000 for cobalt and $60,000 for nickel. This Act expired in April, 1917. The table given below shows the total amount paid in bounties.

Summary of Bounties Paid

Company	Cobalt	Nickel	Total
	$ c.	$ c.	$ c.
Deloro Smelting and Refining Co., Ltd.....	48,930 93	8,166 96	57,097 89
Coniagas Reduction Co., Ltd.	67,174 99	27,539 01	94,714 01
Metals Chemical Ltd.....................	9,577 60	6,766 04	16,343 65
Canadian Smelting and Refining Co., Ltd..	1,026 05	681 84	1,707 89
Standard Smelting and Refining Co., Ltd..	214 92	....	214 92
Dominion Refineries Limited	62 59	....	62 59
Total......................	126,987 08	43,153 85	170,140 95

Previous to 1913 a large amount of cobalt ore had beeen shipped to the United States to be refined, as there was an import duty of 25 cents per pound on refined cobalt oxide.   However, in 1913 the charges on cobalt products entering the United States were changed and at present are as follows:

Product.	Old Tariff	New Tariff
Cobalt oxide	25 cents per lb.	10 cents per lb.
Cobalt metal	free	free
Cobalt ore	free	free
Cobalt alloy steel	0.2 to 7 cents per lb. according to the value and 20 per cent. above 40 cents per lb.	15 per cent. ad. val.

A synopsis follows of the methods employed in Europe for the production of cobalt oxide, and of the numerous processes which have been proposed, many of which have never been put to the test of actual practice.

Herrenschmidt and Constable Processes[1]

These processes were used chiefly on the ores from New Caledonia and Australia.

Process No. 1.—The cobalt ore was smelted with argentiferous lead or copper sulphides in a blast furnace, the products being a matte containing the cobalt, nickel, copper, or lead as sulphides, and a silicate slag containing the iron and manganese. The matte was ground and heated to convert the sulphides of the metals into soluble sulphates, the lead sulphide, however, forming an insoluble sulphate. The sulphate solution was then treated by one of the following methods:

(a) The copper in the solution was precipitated by iron, the solution filtered upon a layer of New Caledonia mineral, the iron in solution precipitating, and a corresponding amount of cobalt, nickel, and manganese dissolving. Magnesia (MgO) was added next to precipitate the nickel, cobalt, and manganese. The precipitate was treated with a fresh quantity of solution containing cobalt, nickel, and manganese sulphates, by which treatment the manganese hydroxide dissolved, while a corresponding quantity of cobalt and nickel hydroxides was precipitated.

(b) To the sulphate solution, calcium chloride was added to change the sulphates into chlorides. The solution was filtered to remove the calcium sulphate and any insoluble matter, and then treated with lime, the iron and part of the copper being precipitated as hydroxides. The last of the copper was precipitated with nickel hydrate or carbonate. The solution was then treated with calcium or sodium carbonate which precipitated the cobalt, nickel, and copper. The precipitate of mixed carbonates or hydrates was redissolved, and the solution treated to remove the copper as above, so that only cobalt and nickel remained in solution.

Process No. 2.—The ore was treated with hydrochloric acid, and the solution of chlorides was reduced by filtering on scrap iron. The solution of ferrous chloride obtained was used to attack fresh quantities of New Caledonia ore. The acid treatment was applied chiefly to the cobalt ores from Australia.

[1] Wagner's Jahresbericht, Vol. XXX, 1884; pp. 152, 396; Vol. XXXII, 1886, pp. 157-158; Vol. XXXVIII, 1892, p. 208; Jour. Soc. Chem. Ind., Vol. XI, 1892, p. 167; Schnabel, Handbook of Metallurgy, Vol. II, 1898, p. 603.

At the plant of the Malètra Co., Rouen, France, the pulverized ore from New Caledonia was treated with a ferrous sulphate solution and steam. The solution after such treatment contained the sulphates of cobalt, nickel, and manganese, the iron forming a basic sulphate precipitate. The iron precipitate, silica, and alumina were removed by filtering and washed in a filter press. To the sulphate solution sodium sulphide was added, which precipitated the sulphides of cobalt and nickel, as well as a small quantity of manganese.

The precipitate was filtered in a filter press, and, after washing, was digested with a calculated quantity of ferric chloride, which dissolved only the manganese sulphide. A black precipitate of cobalt and nickel sulphides remained, the solution containing the sulphates and chlorides of manganese and iron. The sulphide precipitate was dried, ground, and heated to form sulphates. The mixture of sulphates was dissolved in water and treated with a solution of calcium chloride, the calcium sulphate formed being removed. The solution was divided into two parts, A and B. To the first portion of solution A milk of lime was added to precipitate the cobalt and nickel; afterwards the precipitate was washed in a filter press to remove any excess of calcium chloride. The washed precipitate was digested with water and oxidized by chlorine mixed with air. A part of solution B containing cobalt and nickel chlorides was added to the hydroxides obtained from solution A, and the mixture agitated. By this treatment the nickel hydroxide was reduced by the cobalt chloride of solution B and dissolved, a corresponding quantity of cobalt precipitating, according to the following reaction.

$$Co_2O_3 + Ni_2O_3 + 2CoCl_2 = 2NiCl_2 + 2Co_2O_3.$$

Another method of treating the ores from New Caledonia was to make a matte which was afterwards ground and heated to form sulphates. The sulphates were dissolved, and the iron was precipitated by careful additions of sodium carbonate. The copper was precipitated by hydrogen sulphide, and after boiling off the excess, the cobalt was precipitated by sodium hypochlorite. Bleaching powder cannot be used owing to the sulphates in solution. The nickel was precipitated from the remaining solution by sodium carbonate. Also metallic copper was precipitated from copper-cobalt-nickel-iron solutions by the addition of matte. The iron was precipitated by adding a salt of nickel or cobalt.

Borchers and Warlimont[1] proposed treating minerals containing sulphides of nickel, cobalt, copper, and iron by a partial roast at 500° C. In this treatment the cobalt and copper were changed almost completely to sulphates, the nickel remained partly as sulphide and partly as oxide, the iron as oxide and ferrous and ferric sulphates. The furnace product was treated with water for several days, when any copper sulphide remaining in the ore reacted with the ferric sulphate, forming copper sulphate and ferrous sulphate. This was followed by a treatment with a slightly acidulated solution which dissolved the sulphates of cobalt and copper and at the same time a small quantity of nickel and a little iron sulphate. The residue was treated to recover the nickel. The copper was removed from the cobalt solution by scrap iron, and calcium sulphide was added to precipitate the nickel and cobalt as sulphides.

[1] Eug. Prost, Cours de Métallurgie des Metaux autres que le Fer, 1912, p. 663, Ch. Béranger, Editeur, Paris.

Lundborg[1] describes the treatment of ores containing earthy cobalt formerly used at the Editha Smalt Works in Silesia. The ores were agitated with concentrated hydrochloric acid in clay vessels to dissolve the cobalt, some nickel and iron dissolving at the same time. The iron was precipitated from the solution by adding small quantities of marble. As soon as nickel began to be precipitated the separation of iron was known to be complete. To the filtrate soda was added to precipitate the nickel. When cobalt began to come down, the liquid was filtered, and the precipitation continued, a mixture of the two hydroxides forming until all the nickel had been precipitated. Cobalt hydroxide was precipitated from the final filtrate. The precipitate of mixed oxides was redissolved, and the two separated by fractional precipitation.

The matte, from the Sesia Works at Oberschlema, in Saxony, was similarly treated. It contained Ni 16, Co 14, Cu 50, and S 20 per cent. The powdered matte was roasted in a reverberatory furnace and treated with dilute sulphuric acid. The copper was precipitated by iron, after which the rest of the process was the same as that described above.

At Schladming, Styria,[2] ore containing 11 per cent. of nickel and 1 per cent. of cobalt was roasted in stalls and afterwards smelted in blast furnaces. The charge consisted of 89 parts of roasted ore, and 19 parts of quartz. In 24 hours five tons of ore were treated, using 1,800 pounds of wood charcoal. The speiss, which was tapped every two hours, contained Ni 45 to 47, Co 4 to 6, Fe 8 to 10, Cu 1 to 1.5, As 33 to 36, S 1 to 2, and carbon 1 to 2 per cent.

At the George Works at Dobschau, Hungary,[3] about 1876, ore containing nickel 4.5 and cobalt 1.5 per cent. was roasted in stalls, and then smelted in circular blast furnaces 16.5 feet high, with two tuyeres. The diameter of the furnace was 3 feet 4 inches at the level of the tuyere, and 4 feet at the mouth. The tuyeres were 2.75 inches in diameter, and the blast pressure equal to 2.5 inches of mercury. The charge consisted of 100 parts of ore, 3 to 4 quartz, 8 to 12 limestone, and 5 to 10 rich slag. The capacity of the furnace was 7 to 10 tons of ore per 24 hours, using 20 per cent. charcoal. The speiss contained from 16 to 20 per cent. of nickel and cobalt. This was roasted in stalls, and then smelted to form a concentrated speiss in a blast furnace similar to that used for the ore. The capacity of the furnace was about 11 tons per 24 hours. To the charge of roasted speiss, about 2.5 tons of quartz were added to flux the iron oxide formed during the smelting operation. Charcoal was used as a fuel, the consumption being about 26 per cent. of the weight of the charge. The concentrated speiss contained Ni and Co 31.9, Cu 1.9, Fe 26.4, As 36.3, and S 3.1 per cent.

The concentrated speiss was roasted and charged into a small blast furnace similar to that used for the ore. The capacity of the furnace was about two tons of roasted speiss. After the charge (100 parts of roasted speiss, 4 parts of quartz, 2 parts of glass and 1 part of soda), had melted, a strong air-blast was turned on, by which the iron was oxidized first. The iron oxide was slagged by addition of quartz, glass, and soda. To prevent large quantities of cobalt being slagged, the

[2] Schnabel, Handbook of Metallurgy, Vol. II, 1898, p. 602.
[2] Schnabel, Vol. II, p. 562.
[3] Schnabel, pp. 562, 586.

iron was not completely oxidized, 8 to 10 per cent. remaining in the speiss. The speiss formed by this operation was called doubly concentrated speiss and contained Ni and Co 50 to 52, Cu 1 to 2, Fe 8 to 10, As 38 to 40, and S 1 to 2 per cent. The slag contained 1 to 2 per cent. Ni and Co, and was re-charged into the blast furnace treating the ore.

The high-grade speiss was crushed in a stamp-battery, and then roasted for 12 to 14 hours in wood-fired reverberatory furnaces with a capacity of about 600 pounds to a charge. After roasting, 65 to 90 pounds of sawdust or coal dust was added which reduced any arsenates to arsenites, from which compound the last of the arsenic may be removed. The roasted product was treated with sulphuric acid and the solution filtered. Iron and copper were precipitated by additions of calcium carbonate to the boiling solution, after which cobalt hydroxide was precipitated by calcium hypochlorite, and nickel by milk of lime. The cobalt and nickel oxides were dried, washed, ground, and sold to the colour works in Saxony.

At Saint Benoit, near Liege,[1] speiss containing 45 per cent. of nickel was treated with concentrateed hydrochloric acid at 80° C. Iron was precipitated from the solution in the usual way, and then copper by calcium sulphide. Next cobalt was precipitated by chloride of lime, and lastly nickel by milk of lime.

Dixon[2] proposed smelting garnerite, a hydrated nickel magnesium silicate of New Caledonia, with the addition of arsenical materials, to form a speiss. This was roasted, and treated with hydrochloric acid. Into the chloride solution chlorine was passed to oxidize the iron, which was afterwards precipitated by the careful addition of nickel hydroxide. Cobalt was then precipitated as hydroxide by passing more chlorine through the solution, and adding more nickel hydroxide. The solution containing the nickel chloride was evaporated. The salts obtained were heated in a furnace through which steam or hydrogen was passed, producing nickel oxide or metallic nickel respectively. A little nickel was precipitated with the cobalt, but this was removed by leaching with dilute hydrochloric acid. It is also stated that the addition of manganese dioxide to a neutral cobalt and nickel solution completely precipitated the iron, but this is questionable, if not impossible. The statement is also made that anhydrous nickel oxide was used where hydroxide is mentioned above for precipitating iron and cobalt, but knowing the solubility of nickel oxide, the writer also questions such a statement.

At a works in Birmingham, England,[3] speiss was roasted and dissolved in hydrochloric acid. Iron was first oxidized, then precipitated as arsenate and hydrate, by neutralizing the hot solution with calcium carbonate. Copper was precipitated next by sulphuretted hydrogen, then the cobalt by calcium hypochlorite (bleaching powder), and finally the nickel by milk of lime. In precipitating the cobalt a small amount of quicklime was added to neutralize any liberated acid.

A direct treatment of unroasted matte with acid was formerly in use at the Scopello works in Piedmont.[4] Nickel matte, containing Ni 24, Co 6, Cu 12, Fe 23, and S 35 per cent. was treated with hydrochloric acid (33 per cent. HCl) in stoneware vessels surrounded by water. The sulphuretted hydrogen formed was

[1] Schnabel, Vol. II, p. 586.
[2] Chemical News, Vol. XXXVIII, 1878. pp. 268-270.
[3] Phillips. Elements of Metallurgy, 1891, p. 415.
[4] Schnabel, Vol. II, p. 581.

removed by a tube in the cover of the vessel and burned. After the matte had been treated three times with acid, the liquid was removed from the residue, which consisted of the copper sulphide of the matte and an appreciable quantity of nickel and cobalt sulphides. This was charged into the blast furnace during the smelting of matte or ore. The solution, containing chlorides of iron, nickel, and cobalt, was allowed to settle, and then evaporated to dryness in a cast-iron pot. The residue, a mixture of the three chlorides, was heated in a reverberatory furnace for 3 or 4 hours, with continual rabbling, during which process part of the iron was volatilized as chloride and part changed to ferric oxide.

The furnace product was agitated with water to dissolve the soluble cobalt and nickel chlorides, any undecomposed iron chloride present dissolving at the same time. The iron chloride was oxidized with chloride of lime and precipitated with calcium carbonate. Cobalt was precipitated next by further additions of chloride of lime to the iron-free solution, and finally nickel was precipitated with milk of lime.

The precipitates of cobalt and nickel hydroxides were washed in woollen sacks to remove any soluble lime salts, until no cloudiness was visible in the water on the addition of ammonium oxalate. The oxides were given a final wash with acidulated water.

Other Proposed Processes

Beltzer[1] outlines a process for the treatment of ores from Cobalt, Canada, as follows. The ore was to be first concentrated and the silver removed by amalgamation. The concentrate, after amalgamation, is roasted (a) with or without additions of carbon to remove the arsenic, as oxide, or (b) treated with lime or soda and the arsenic removed as soluble arsenite or arsenate, or (c) heated with sodium bisulphate and acid (66° Bé.) to remove the last traces of arsenic. By the last method of treatment the soluble sulphates of cobalt and nickel are formed. In case of treatment (a) or (b), the roasted product is afterwards given either a chloridizing roast with salt, or heated with hydrochloric acid, or sulphuric acid and salt. In any case, after dissolving the soluble salts and filtering the solution containing chlorides of nickel, cobalt, and a small quantity of iron, the cobalt is precipitated by Rose's method (caustic soda and current of chlorine) or by hypochlorite of calcium or sodium. The precipitation of cobalt may also be completed by the following method. If the cobalt and nickel are in the form of sulphates, calcium chloride is added, which converts the sulphates into chlorides. The solution is filtered and then divided into two parts, A. and B. The cobalt and nickel in solution A are completely precipitated by milk of lime, filtered and washed. The precipitate of cobalt and nickel hydroxides is mixed with water and oxidized by a current of chlorine and air. When the solution is saturated with chlorine, a part of solution B is added and the mixed solution boiled. The nickel hydroxide is reduced by the cobalt chloride of solution B and dissolves, a corresponding quantity of cobalt precipitating. In this way practically pure cobalt hydroxide is obtained.

[1] Beltzer, The Rational and Industrial Treatment of the Complex Ores of Silver, Cobalt, Nickel and Arsenic of Cobalt, Canada. Moniteur Scientifique, Vol. XXIII, 1909, pp. 633-647.

Quantities of solution B are added until the precipitate is practically pure cobalt hydroxide.

Beltzer also outlines the following electrolytic process to separate cobalt from nickel, but the process does not appear to be practicable. To the chloride solution salt is added and the solution electrolyzed, using platinum electrodes. Nascent chlorine and sodium hydrate are formed, the caustic soda precipitating the cobalt, the nickel remaining in solution.

A solution containing sulphates of cobalt, nickel, and a small quantity of iron may be treated by the following method. The iron is precipitated by milk of lime or nickel hydrate, a corresponding quantity of lime or nickel going into solution. Soda is added to precipitate the cobalt and nickel which are separated by additions of chlorine to a neutral solution. In the oxidation with chlorine, nickel dissolves, and at the same time a corresponding quantity of cobalt is precipitated so that the original rose colour, due to the cobalt, changes to a green. To be certain that the cobalt hydrate precipitate does not contain nickel, it is necessary to leave a little cobalt in solution.

Another process[1] consisted in mixing cobalt-silver ore with a lead furnace-charge and smelting. The lead collected the silver, and the cobalt and nickel formed a speiss. The speiss was roasted with carbon to remove arsenic, the cobalt combining with silica, which was added as silicate. The cobalt silicate was treated with hydrochloric acid, and the cobalt and nickel hydroxides were precipitated by lime water.

It does not appear to the author that the processes to treat cobalt ores, as outlined on the following pages, with perhaps a few exceptions, are anything more than laboratory experiments or ideas. As mentioned in the introduction, a summary of each is given merely to record all attempts that have been proposed to treat cobalt ores.

Readman[2] proposed mixing the ground ore with sodium sulphate in sulphuric acid, in the proportions of 100: 84: 65. The mixture was heated to a red heat, the cobalt, nickel, and manganese forming soluble sulphates. After dissolving the sulphates, the solution was neutralized with calcium carbonate and heated, the iron precipitating. The sulphate solution of cobalt, nickel, and manganese was treated with sodium sulphide to precipitate only the cobalt and nickel. In another process the use of ferric chloride was suggested. After treating the ore, the product was heated to render the iron insoluble. The cobalt and nickel were dissolved from the roasted mass.

Gauthier[3] used iron pyrites and gypsum to decompose cobalt ore, obtaining sulphates of cobalt and nickel.

Stahl[4] proposed roasting the mineral, then mixing it with salt and sulphide of iron. The copper, cobalt, and nickel were changed to soluble chlorides, the iron and manganese to insoluble oxides. The chloride solution was treated with

[1] Ouvrard, Industries du Chrome, du Manganese, du Nickel, et du Cobalt, p. 258, Doin and Sons, Paris, 1910.
[2] Ouvrard, p. 262; Jour. Soc. Chem. Ind., Vol. III, 1884, p. 524.
[3] Wagner, Jahresbericht, Vol. XXXII, 1886, p. 157.
[4] German Patent, No. 58,417, 1890. Wagner, Jahresbericht, Vol. XXXVII, 1891, p. 210, Ouvrard, pp. 265-266. Schnabel, Vol. 2, p. 605.

hydrogen sulphide to precipitate the copper, and after neutralization with soda, cobalt and nickel were precipitated by sodium sulphide along with any iron, manganese, and copper. The mixed sulphides were treated with dilute acid to dissolve any iron, manganese, or copper sulphides. The cobalt and nickel sulphides were roasted to sulphates and separated by one of the previously mentioned methods.

In the Natusch or Schoneis process[1] the roasted mineral was heated with ferric chloride. The chloride solution was treated with calcium sulphide, which precipitated the cobalt and nickel and a small quantity of manganese as sulphides.

Precourt and Falliet[2] suggested roasting cobalt ore with iron sulphide, FeS_2, the cobalt, nickel, and manganese being changed to soluble sulphates, most of the iron forming oxide. The iron was precipitated from the solution with calcium carbonate and removed. The sulphates of cobalt, nickel, and manganese were changed to chlorides by the addition of calcium chloride, calcium sulphate being precipitated and removed. Nickel and cobalt were precipitated with calcium sulphide, the manganese remaining in solution. The sulphides of nickel and cobalt were washed with dilute sulphuric acid to remove any manganese sulphide, and afterwards heated to form sulphates. To the filtered solution of the sulphates, hypochlorite was added. After making the solution alkaline it was made faintly acid, the cobalt precipitating as sesquioxide, the nickel remaining in solution.

Dyckerhoff[3] proposed decomposing ore containing silver, cobalt, nickel, and arsenic by roasting the ore with salt, clay, and pyrites, by which treatment the silver, cobalt, nickel, and arsenic were converted to chlorides. The insoluble silver chloride was removed, while the arsenic chloride was volatilized or changed to the volatile oxide. The cobalt and nickel remained as soluble chlorides.

In the Warren process,[4] cobalt ore was treated with hydrochloric acid and copper nitrate, the metals passing into solution. Milk of lime was added, which precipitated the iron as hydroxide and arsenate. Lime was removed by the addition of sulphuric acid, and afterwards the cobalt and nickel were precipitated as carbonates by soda. The solution was filtered, diluted, and chlorine gas passed through until the solution was saturated, after which it was boiled, the nickel hydroxide dissolving and cobalt hydroxide remaining. The solution was filtered and the nickel precipitated by caustic soda.

Gauthier[5] heated ground cobalt ore with hydrochloric and sulphuric acids and completed the process by the usual methods.

Carnott[6] heated the ore to render any iron insoluble, and then treated the product with hydrochloric acid and neutralized the solution with calcium carbonate to precipitate the iron. The filtered solution was treated with milk of lime, which precipitated first the cobalt, then the nickel, and finally the manganese. The fractional precipitation of the cobalt, nickel, and manganese was not satisfactory.

[1] Ouvrard, p. 264. Schoneis, Berg. u hüttenm. Zeitung, 1890, p. 453. Wagner, Jahresbericht, Vol. XXXVI, 1890, p. 338. Chemiker Zeitung, Vol. XIV, 1890, p. 770, p. 1475.
[2] French Patent 403,830, June 9, 1909. Jour. Soc. Chem. Ind., Vol. XXIX, 1910, p. 97.
[3] United States Patent. No. 1,085,675, Feb. 3, 1914.
[4] Chemical News, Vol. LVI, 1887, p. 193.
[5] Gauthier, Wagner Jahresbericht, Vol. XXXII, 1886, p. 157.
[6] Ouvrard, p. 258.

Clark[1] treated the ground mineral with a boiling solution of ferric chloride. The solution was evaporated to dryness and the residue calcined at 350° to 370° C. Cobalt, nickel, and manganese were changed to chlorides by the decomposition of the ferric chloride which formed ferric oxide. Cobalt and nickel sulphides were separated from manganese by precipitation with calcium sulphide. This method was tried at Glasgow, Scotland.

Dickson and Ratte[2] dissolved the cobalt, nickel, and manganese of a cobalt mineral by treatment with sulphurous acid, alone, or mixed with other acids. The iron oxide remained insoluble. The finely-ground cobalt mineral was mixed with water at 50° C. in vats provided with stirring arrangements, and the sulphurous acid introduced. The mixture was transferred to a second vat, where the iron oxide and the impurities settled. The solution was filtered through a bed of pulverized mineral, which retained the suspended impurities and neutralized the excess acid, while the last traces of iron were precipitated. From this solution the cobalt and nickel were precipitated as sulphides by sodium sulphide and afterwards separated.

Cobaltite was ground and carefully roasted with additions of small quantities of carbon.[3] The residue was treated with sulphuric acid. From the solution obtained the iron was precipitated with calcium carbonate, and other metals with hydrogen sulphide. The solution was filtered and treated to recover the cobalt and nickel.

Barth[4] gives some experimental results obtained in working on a method to treat a roasted speiss containing lead, cobalt, nickel, iron, and manganese. Decomposition by acids, chlorine, chloridizing roasting, and by sulphur dioxide was tried.

Phillips[5] outlines a process for smelting cobalt silver ores using sufficient iron matte and copper residue to form a speiss, an argentiferous copper matte, and a slag. The operation was repeated to obtain speiss free from silver and copper.

De Burlet[6] proposed extracting cobalt from cobalt and nickel silicate minerals and slags from copper smelting by treating them with dilute sulphuric acid, and after decomposition the mass was heated to 150°-200° C. to render the silica insoluble. The soluble salts were dissolved in hot water, the solution filtered, and calcium carbonate added to precipitate the iron and the greater part of the copper. The remaining copper was removed by electrolysis, and then the solution was treated with ammonia to obtain cobalt and nickel compounds. The ammoniacal solution was electrolyzed, using carbon or lead anodes and polished nickel-plated iron plates as cathodes. The surfaces of the cathodes were coated with paraffin to prevent the metal adhering. The metal fell to the bottom of the tank in thin flakes.

---

[1] Ouvrard, p. 263.

[2] French Patent, Sept. 5th, 1885.

[3] Ouvrard, p. 258.

[4] Barth, Treatment of a Roasted Lead-Cobalt-Nickel Speiss, Metallurgie, Vol. IX, 1912, p. 199. Chem. Abst., Vol. VI, 1912, p. 1732. Jour. Soc. Chem. Ind., Vol. XXXI, 1912, p. 391.

[5] United States Patent, 1,127,506, 1915.

[6] British Patent, 27,150, Nov. 25, 1913.

Cito,[1] after experimenting with other processes, patented the following treatment for ores from Cobalt, Canada. The raw ore was treated in a reverberatory furnace with copper and fluxes. Two products were obtained, viz., an alloy containing the copper and all the silver, nickel, cobalt, and arsenic; and a slag. The alloy was cast from the furnace into anode moulds. The metals were recovered by electrolysis, using an electrolyte of copper sulphate, and sheet copper cathodes. The copper was deposited on the cathodes in a pure form, the silver precipitated as slime, and the cobalt and nickel remained in solution. The arsenic was found partly in solution and partly as slime.

From the electrolyte containing copper, cobalt, nickel, and arsenic, copper was precipitated in the cold solution, and arsenic later in the hot solution, by hydrogen sulphide. The cobalt nickel solution was treated to recover the cobalt and nickel by the ordinary processes.

Andre[2] attempted to recover cobalt and nickel from a cobalt-nickel matte as an anode and using an electrolyte of dilute sulphuric acid or ammoniacal cobalt sulphate. A frame containing granulated metal was placed between the anode and cathode and on this copper and silver were precipitated.

Vortmann[3] proposed a process to obtain by electrolysis cobaltic oxide or hydrate from solutions containing cobalt and nickel. This process was based on the assumption that if a current is passed through solutions of cobalt and nickel containing no alkaline sulphates or other neutral salts of the alkalies, cobaltous and nickelous hydrates or basic salts of both form at the cathode. If the current is reversed the nickelous hydrate or corresponding basic salt dissolves, while cobaltous hydrate is oxidized to cobaltic hydrate. On changing the current to its original direction more of each lower hydrate is produced, and on again reversing the current the nickel is dissolved. In this way all the cobalt is finally precipitated as hydrate, and all the nickel remains in solution. If there is a small quantity of a chloride present in the liquid (equivalent to 1 per cent. of common salt), the cobaltous hydrate is very quickly oxidized to the higher compound by the small amount of chlorine or hypochlorous acid set free. In this case the constant change of current is unnecessary.

The separation of cobalt is assisted by gentle warming. After the precipitation is completed, the current is stopped and the liquid heated to 60° or 70° C. whereby any small quantity of nickelic hydrate remaining in the cobalt compound is dissolved. The nickel solution when filtered does not contain any cobalt.

Guiterman[4] has patented a process to extract cobalt oxide by electrolyzing a nickel and cobalt solution containing chlorides. The chlorine liberated at the anode reacts with the electrolyte to form hypochlorite, the hypochlorite reacting with the cobalt in solution to form hydrated oxide of cobalt and some free hydrochloric or sulphuric acids. To prevent the electrolyte becoming too acid, in which case the cobalt hydrate would redissolve, sodium carbonate solution is added.

[1] Cito, Trans. Amer. Electrochemical Soc., Vol. 17, 1910, p. 239.
[2] Ouvrard, p. 265
[3] German Patent, No. 78,236, May 10, 1894, Schnabel, Vol. II, p. 608.
[4] United States Patent 1,195,211, August 22nd, 1916.

The firm of Basse and Selve, Altena, Germany, have suggested a process[1] which consists first in adding certain organic salts to neutral or slightly acid solutions containing nickel, cobalt, iron and zinc, such as will prevent the precipitation of their oxides by alkalies. Such are acetic acid, citric acid, glycerine, and dextrose. The solution is made alkaline by soda or potash lye, and subjected to electrolysis with a current of 2.8 to 9.3 amperes per square foot. Iron, cobalt and zinc are deposited on the cathode, while nickel either remains entirely in the liquid or precipitates partly as hydrate, according to the alkalinity of the solution. The precipitation of the hydrate occurs if the current is continued for a long time. To the nickel solution free from other metals, ammonium carbonate is added to form carbonate of all the free alkali, after which it is electrolyzed. Nickel is formed as a bright deposit on the cathode.

Coppet[2] smelted ore to obtain a matte, which was ground and roasted to form oxides of copper, nickel, and cobalt. The oxides were reduced to the metallic state, then treated with a solution of a cupric salt. The copper was precipitated afterwards by metallic cobalt and nickel. No further explanation is given of this process. In another process the same author roasted matte to form sulphates or chlorides by the addition of salt. Copper was removed as above.

To treat nickel ores or products containing cobalt, the following methods have also been suggested. The complete treatment is not given in all cases but only those parts which differ from other processes.

Laugier[3] dissolved the ore in nitric acid, and evaporated the solution, the arsenic being precipitated as oxide during the evaporation or later by hydrogen sulphide. The solution was filtered, and heated till the excess of sulphuretted hydrogen was expelled, then the iron was oxidized. Sodium carbonate was added in excess, while the liquor was hot, to precipitate the nickel and cobalt in the form of carbonates, and the iron as hydrate. The precipitate was then washed and digested with an excess of oxalic acid solution, the soluble ferric oxalate being separated by filtration from the oxalates of nickel and cobalt which are insoluble even in excess of oxalic acid. The latter salts were mixed with dilute ammonia in a closed vessel. (Stromeyer[4] recommends strong ammonia), to dissolve the oxalates. The filtered solution, after exposure to the air for several days, deposited nickel ammonium oxalate and manganese oxalate, while the pure oxalate of cobalt remained in solution. The oxalate of nickel separated as above can be freed from the small quantity of the cobalt salt which precipitates with it, by re-treatment with ammonia. The residue, obtained by evaporating the ammoniacal solution of the cobalt oxalate, yielded sesquioxide of cobalt when ignited in the air, or metallic cobalt when ignited out of contact with the air.

Guesneville[5] treated cobalt ore with nitric acid, the iron and arsenic being removed as ferric arsenate by the addition of potassium carbonate. The cobalt was precipitated as oxalate by potassium acid oxalate.

---

[1] Schnabel, p. 591.

[2] Ouvrard, p. 261. Jour. Soc. Chem. Ind., Vol. XII, 1893, p. 274.

[3] Laugier, Annales de Chemie et de Physique, Paris, Vol. IX, 1818, p. 698.

[4] Stromeyer, Jour pract. Chemie, Vol. LXVII, 1856, p. 185.

[5] Guesneville, Gmelin Kraut, Handbuch der anorganischen Chemie, Band 5, 1, 1909, p. 192.

De Witt[1] treated speiss with aqua regia and removed the excess acid. Ammonium chloride and ammonia were then added, followed by additions of potassium acid oxalate. The cobalt was recovered as oxalate.

Loujet[2] dissolved cobalt ore with hydrochloric acid, and to·the solution a ferric salt was added, followed by additions of potassium carbonate, calcium carbonate, or calcium hydroxide. The iron and arsenic were precipitated as an insoluble ferric arsenate.

Wöhler[3] fused cobalt speiss with potassium carbonate and sulphur in the ratio 1: 3: 3. Most of the metals were converted into simple sulphides, but the arsenic sulphide formed a soluble potassium sulphoarsenate. The temperature had to be regulated so that the cobalt sulphide did not fuse, since it would enclose portions of the soluble sulphoarsenate.

Hermkstädt[4] fused cobalt glanco with potassium nitrate, and the arsenic was removed as potassium arsenate.

Patera[5] roasted ore with additions of carbon, then fused the product with calcium nitrate, soda, and potassium nitrate. Arsenic was removed as soluble arsenates of calcium, soda, and potassium. Nickel was precipitated from the solution by potassium or ammonium bisulphate as a difficultly-soluble complex salt containing only a small quantity of the cobalt compound.

Liebig[6] roasted the ore, then mixed it with ferrous sulphate and potassium bisulphate, and fused the mixture. The arsenic was removed as insoluble ferric arsenate. Ferrous sulphate was added before fusion to prevent the formation of cobalt arsenate. During the final stages of the fusion, the nickel sulphate was decomposed, forming oxide which did not dissolve with the cobalt. Cobalt carbonate was afterwards precipitated by the addition of potassium carbonate.

Barton and McGhie[7] suggested fusing arsenical minerals with sufficient sodium carbonate to combine with the arsenic to form soluble sodium arsenate.

McKenna[8] fused cobalt and nickel speiss with boric acid about equal in weight to the cobalt in the speiss. The heavier speiss separates from the lighter cobalt borate slag.

To separate cobalt from nickel, iron and manganese, the following processes have been suggested:

Sack,[9] by the addition of lead peroxide to a solution of cobalt, manganese, aluminum, and iron salts, precipitated hydrates of manganese and aluminum, basic ferric-sulphate and lead sulphate, practically all the cobalt remaining in solution. Any copper was first removed, and then the solution was mixed with a calculated amount of peroxide of lead. In case iron was present in large quantities, it was

---

[1] Jour. prakt. Chemie, Leipzig, Vol. 71, 1857, p. 239.
[2] Loujet, Monit. Indust., 1849, p. 1309. Jour. Soc. Chem. Ind., Vol. 1, 1882, pp. 258-259.
[3] Wöhler, Annalen der Physik and Chemie, Vol. VI, 1826, p. 227.
[4] Hermkstädt, Jour. für Chemie und Physik, Vol. XXXI, 1821, p. 105.
[5] Patera, Jour. prakt. Chemie, Vol. XVIII, 1830, p. 164.
[6] Liebig, Annalen der Physik und Chemie, Vol. XVIII, 1830, p. 164.
[7] French Patent, No. 387,766, Jan. 28th, 1908.
[8] United States Patent No. 1,166,067, Dec. 28th, 1915.
[9] Sack, German Patent No. 72,579, 1892.

precipitated after the copper with an alkaline or alkaline-earth carbonate. If a large amount of manganese was present, it was removed by fractional precipitation with a soluble alkaline or alkaline-earth sulphide.

Iron was precipitated from a cobalt nickel solution by additions of cobalt hydrate.[1]

To a solution containing cobalt, nickel and iron, soda was added which precipitated the iron, then ammonium chloride was added followed by potassium hydrate, and the mixture heated. Nickel hydroxide was precipitated, the precipitation of cobalt hydroxide being proportional to the decomposition of the ammonium chloride.[2]

Solutions containing cobalt, nickel, iron and manganese were treated with sodium acetate and heated, the iron being precipitated. Cobalt was precipitated from the neutral solution by hydrogen sulphide, the manganese acetate not being decomposed. Before the treatment with sodium acetate, the copper and arsenic were removed with hydrogen sulphide in an acid solution. Precipitation of the cobalt may also be made by the addition of potassium or barium sulphide. The sulphide precipitate was washed with cold dilute hydrochloric acid, which removed the sulphides of manganese, zinc, and iron, the cobalt sulphide remaining undissolved.[3]

A neutral solution of cobalt and nickel is mixed with potassium nitrite, potassium cobalt nitrite being formed. The presence of lime interferes with this separation, as a potassium lime nickel nitrite is precipitated at the same time.[4]

Barton and McGhie[5] separate cobalt and nickel from chloride solutions by subjecting the slightly acidified solution to fractional crystallization.

Miscellaneous Processes Summarized

Hybinette[6] outlines a process of separating copper from cobalt and nickel. The ore is roasted and the product separated by magnetic concentration. The magnetic concentrate, containing copper, cobalt, and nickel, is smelted to form a matte, which is roasted and leached with dilute sulphuric acid. A solution containing principally copper, and only small amounts of iron and cobalt, is obtained.

It has been proposed[7] to produce from sulphide ores a matte free from iron, and containing only sulphur, nickel, and cobalt. This matte was fused under a blast on a bed of quartz and sodium silicate in order to produce a silicate of cobalt. The latter compound was fused with soda and nitre to liberate the cobalt oxide. The method, however, does not appear to have come into use.

Richer[8] gives an interesting account of an old method to separate cobalt from nickel by crystallization from a solution containing a large quantity of ammonium sulphate. Mention is also made of the possibility of removing copper from cobalt oxide by heating with ammonium chloride.

---

[1] Loujet, Monit. Indust., 1849, p. 1309.

[2] Phillips Wittstein, Repertorium für die Pharmacie, Vol. 57, p. 226.

[3] Wackenbroder, Gmelin Kraut, Handbuch der anorganischen Chemie, Band V, I, 1909, p. 193.

[4] Stromeyer, Jour. prakt. Chemie, Vol. LXVII, 1856, p. 185.

[5] German Patent 222,231, 1905. Metallurgie, Vol. VII, 1910, pp. 667-674.

[6] United States Patent, No. 1,098,443, June 2nd, 1914.

[7] Schnabel, Vol. II, p. 601.

[8] Tilloch, Phil. Mag., Vol. XIX, 1804, pp. 51-54.

Bücholz[1] outlines a method proposed to prepare cobalt and nickel oxides by dissolving the hydroxides in ammonia after precipitation.

Hauer[2] describes Patera's application of analytical methods for the production of cobalt and nickel oxides. In the process the iron and arsenic were precipitated with powdered calcium carbonate, the cobalt by calcium bleach, and the nickel by milk of lime.

Vivian[3] was granted a patent covering a process based on the affinity cobalt and nickel have for arsenic. In the specifications the claim is made that it is possible to separate cobalt and nickel from copper by regulating the amount of arsenic, and by having some sulphur present to combine with the copper.

Wright[4] proposed extracting cobalt and nickel from waste solutions from copper refining, by adding milk of lime, which precipitated the metals. After drying the precipitate it was mixed with sand, 15 to 20 per cent. of residue from the alkali works, 15 to 20 per cent. of carbon, and, if possible, with products containing arsenic. The cobalt and nickel were recovered as speiss.

Careis[5] proposed the following method to extract cobalt from ores. The ore was first dissolved in hydrochloric acid. The metals in solution were precipitated with soda, and the precipitate dissolved in sulphuric acid. The solution was neutralized with hot dilute sulphuric acid, whereby the copper and other metals, especially iron, were precipitated, the cobalt and nickel remaining in solution. The solution was mixed with hot ammonium or potassium sulphide, which precipitated the nickel. Metallic zinc was added to precipitate the cobalt from sulphate solution.

Grosse-Bohle[6] patented a process to precipitate cobalt and nickel from sulphate and chloride solutions by means of zinc.

Aaron[7] proposed precipitating cobalt and nickel from solutions as methylsulphocarbonates.

Neill[8] described the method used to treat the Mins la Motte ores.

Pelatan[9] published a clear description of the Herrenschmidt process as used at the Malétra works, Rouen, France.

Herrenschmidt and Capelle[10] issued a report for the French Government on the following processes: Carnot, Readman, Herrenschmidt, Clarke, Dixon, and Ratte.

Kripp[11] attempted to recover cobalt and nickel from copper ores containing silver. The silver was changed to chloride and removed, after which operation the

---

[1] Tilloch, Phil. Mag., Vol. XXIII, 1805, pp. 193-199.

[2] Jour. prakt. Chemie, Vol. LXVII, 1856, pp. 14-24.

[3] British Patent No. 13,800, Nov. 4, 1851; Percy, Metallurgy: Fuel, Clays, Copper, Zinc, etc., 1861, pp. 375-378.

[4] Bull. Soc. Chem., Vol. V (1st series), 1866, 475-476.

[5] Berg. u. hüttenm. Zeitung, Vol. XL, No. 23, 1881, pp. 215-216.

[6] German Patent No. 97,114, 1898. Abst. Fischer's Jahresbericht, Vol. XLIV, 1898, pp. 169-170.

[7] United States Patent, No. 330,454, Nov. 17th, 1885.

[8] Trans. Amer. Inst. Min. Eng., Vol. XIII, 1884-1885, pp. 634-639. Eng. Min. Jour., Vol. XXXIX, 1885, pp. 108-109.

[9] Genie Civil, Vol. XVIII, 1891, pp. 373-374.

[10] Moniteur Industriel, Vol. XV, 1888, pp. 145, 156, and 162.

[11] Wagner's Jahresbericht, Vol. XIV, 1868, pp. 111-112.

sulphur was precipitated with barium chloride, the iron by calcium carbonate, and the cobalt by a solution of calcium hypochlorite, stopping the precipitation at a reddish-coloured solution.

Hoepfner[1] was granted a patent covering the following process of treating cobalt ores. A matte was first produced, which was afterwards ground and treated with a cupric chloride solution. By this treatment the sulphides are dissolved, forming cuprous chloride. The metals were removed by electrolysis.

Hanes[2] gave some results obtained from the action of ammonia on cobalt-nickel arsenides. Hydroxides of cobalt and nickel are formed which dissolve in excess of ammonia. The metals may be obtained by electrolysis or by precipitation.

Metals Extraction Corporation[3] patented a process for the extraction and recovery of cobalt and nickel from ores and oxidized mattes. The ore or roasted matte was treated with magnesium chloride solution under pressure. The cobalt dissolved before the nickel.

Pederson[4] gave the results of an investigation of treating cobalt and nickel ores. The article deals more with ordinary chemical reactions.

Borchers[5] gave the following description of a process to treat ores and smelter products containing cobalt, nickel, and silver. The ores or products were first treated with alkaline bisulphate below 200° C., then roasted at 600 to 700°. The product was leached to remove the soluble sulphates. Refractory ores were first roasted with carbon.

Bernard[6] separated cobalt from nickel by precipitating the cobalt by hypo-chlorite that had been previously neutralized and freed from alkaline carbonates and caustic alkalies.

Lance[7] patented a process to separate the hydroxides of copper, zinc, cadmium, silver, nickel, cobalt, and tungsten. The process was based on fractional precipitation by ammonia.

Johnson[8] described a process for the treatment of copper matte containing cobalt, as follows. Matte containing copper 39 per cent., iron 1, cobalt 1, and sulphur 20 per cent., was crushed to 80-mesh and leached with hot 10 per cent. hydrochloric acid. The solution contained cobalt, nickel, and iron. The cobalt and iron were removed by treatment with chlorine and sodium carbonate or by hypochlorites.

Schreiber[9] devised the following process for the separation of cobalt, nickel, and manganese from crude liquors. The iron was precipitated first by the addition

---

[1] British Patent, No. 11,307, 1894. Abst. Jour. Soc. Chem. Ind., Vol. XIV, 1895, p. 754.
[2] Jour. Can. Min. Inst., Vol. VIII, 1905, pp. 358-362.
[3] French Patent, No. 367,717, July 4th, 1906. Jour. Soc. Chem. Ind., Vol. XXV, 1906, p. 1155.
[4] Metallurgie, Vol. VIII, 1911, p. 335. Abst. Jour. Soc. Chem. Ind., Vol. XXX, 1911, p. 900.
[5] British Patent, No. 18,276, August 8th, 1912. Jour. Soc. Chem. Ind., Vol. XXXII, 1913, p. 980.
[6] French Patent, No. 354,941, June 5th, 1905. Jour. Soc. Chem. Ind., Vol. XXIV, 1905, p. 1177.
[7] French Patent No. 342,865. May 3rd, 1904, second edition, Jan. 9th, 1905. Jour. Soc. Chem. Ind., Vol. XXIV, p. 845, 1905.
[8] United States Patent, No. 825,056, July 3rd, 1906.
[9] German Patent, No. 203,310, Sept. 29th, 1907. Jour. Soc. Chem. Ind., Vol. XXVII, 1908, p. 1158.

of calcium carbonate, then the copper by passing in hydrogen sulphide, and finally the cobalt practically free from nickel and manganese was precipitated with calcium hypochlorite. After removing the cobalt the precipitation was continued to precipitate the nickel and manganese, which was dissolved and re-precipitated.

Foote and Smith[1] discuss the dissociation pressures of certain oxides of copper, cobalt, nickel, and antimony.

Chesneau[2] prepared a number of the higher sulphides of cobalt and nickel and determined their solubility.

Mourlot[3] conducted a few experiments to determine the effect of high temperatures on copper, bismuth, silver, tin, nickel, and cobalt sulphides. He found cobalt sulphide is obtained by heating the anhydrous sulphate. At a high temperature it loses all the sulphur, the metal combining with any carbon present.

Manhès[4] claims to have devised an improved process for the treatment of arsenical and sulphide ores of cobalt and nickel. He first produced a speiss or matte which is either dissolved in hydrochloric acid or by electrolysis; or in a second dry refining process, air is blown through the matte which oxidizes the iron and sulphur. The metallic oxides formed by roasting were reduced to metal by carbon and lime, and the metal was used as anodes in the electrolytic refining. In a later patent[5] Manhès suggests adding coke in a converter for blowing matte. He also tried to prepare metallic cobalt and nickel from matte by adding fluxes to remove the sulphur.[6]

Garnier[7] also proposed blowing matte in a converter.

Langguth[8] describes a process used to smelt the cobalt and nickel ores of Norway. The ores were smelted in a blast furnace to produce a matte containing 30 per cent. cobalt and nickel. This was concentrated in a converter to 75 per cent., producing a slag containing 1 to 2 per cent. cobalt and nickel. The blowing operation required 20 to 25 minutes. Reference is also made to Manhès' work.

Savelsburg[9] described a process of blowing nickel and cobalt matte in a converter. Ground matte was blown without the application of heat to oxidize the iron without changing the sulphur content. The product was in a sintered condition satisfactory for melting.

Savelsburg and Papenburg[10] patented a process to convert oxide ores, especially those of cobalt and nickel, into sulphides. Crushed ore was mixed with sulphides and carbon, and briquetted. These were heated in a kiln furnace.

---

[1] Jour Amer. Chem. Soc., Vol. XXX, 1908, pp. 1344-1250.

[2] Comp. Rend., Vol. CXXIII, 1896, pp. 1068-1071. Abst. Jour. Chem. Soc. (London), Vol. LXXII (2), 1897, p. 172.

[3] Comp. Rend., Vol. CXXIV, 1897, pp. 768-771. Abst. Jour. Chem. Soc. (London), Vol. LXXII, 1897, pp. 372-373.

[4] Jour. Soc. Chem. Ind., Vol. IV, 1885, p. 120.

[5] British Patent, No. 17,410, 1888; German Patent, No. 47,444, 1888.

[6] Jour. Soc. Chem. Ind., Vol. XIV, 1895, p. 581. British Patent, 6,914, 1894; German Patent, No. 47,427, 1894.

[7] Eng. and Min. Jour., Vol. XXXVI, 1883, p. 393.

[8] Min. and Sci. Press, Vol. LV, 1888, p. 102.

[9] German Patent, No. 222,231, Jan. 21st, 1908. Abst. Jour. Soc. Chem. Ind., Vol. XXIX, 1910, p. 1257.

[10] German Patent, No. 172,128, Jan. 21st, 1905. Abst. Fischer's Jahresbericht, Vol. LII, 1906, pp. 222-223.

Becquerel[1] applied the electric current to remove cobalt from solutions. The electrolysis was carried on in a cobalt chloride solution that had been neutralized with ammonia or caustic potash. Cobalt was deposited as a brilliant coating. Of the chlorine part escaped and part formed acid. The presence of too much acid gave a dark deposit.

Cohen and Solomon[2] were granted a patent for the electrolytic separation of cobalt from nickel. The addition of strongly oxidizing agents, especially persulphates, cause the precipitation of the cobalt first.

Le Roy[3] described an electrolytic method to extract cobalt and nickel.

Armstrong[4] patented a process for the treatment of complex cobalt ores and for refining cobalt and nickel arsenical and silver-bearing ores. The metals were obtained as chlorides, from which the cobalt was precipitated electrolytically as oxide.

Burlet[5] attempted to extract cobalt, nickel or copper from ores or products as follows. The silicate ore or slag was fused to remove the greater part of the copper as impure metal. The slag was then ground and treated with sulphuric acid. The iron and part of the copper were precipitated with calcium carbonate. The remaining copper, cobalt, and nickel were removed by electrolysis.

Wiggin and Johnstone[6] suggested improvements in the preparation of cobalt and nickel oxides. A cobalt-nickel-copper solution was electrolyzed, using a copper or brass cathode and a carbon anode. The copper was deposited, leaving in solution the cobalt and nickel, which were recovered in the ordinary way.

Martin[7] reduced cobalt and nickel as well as other metals in the ores by passing hydrocarbons over the ore at a bright red heat. The metals were afterwards recovered as an alloy by melting.

Mindeleff[8] attempted to extract cobalt and nickel from ores by reducing the compounds with hydrocarbons, and removing the metals obtained with a magnet.

Berndorfer Manufacturing Company[9] produced malleable and ductile nickel and cobalt by mixing the powdered metal with potassium permanganate, maximum 4 per cent., and melting the mixture.

Selve and Lotter[10] obtained nickel and cobalt free from oxides by the addition of 1.5 per cent. manganese during melting.

Krupp[11] found that cobalt and nickel were not malleable owing to the presence of carbon, but that the defect could be removed by adding manganate or permanganate of potash.

---

[1] Comp. Rend., Vol. LVIII, 1862, pp. 18-20.
[2] German Patent, No. 110,615, 1900. Fischer's Jahresbericht, Vol. XLVI, 1900, p. 157.
[3] Bull. Soc. Industrielle de Mulhouse, Procès-verbeaux, Vol. LXXI, 1901, pp. 154-155.
[4] United States Patent, No. 881,527, March 10th, 1908.
[5] British Patent, No. 27,150, Nov. 25th, 1913. Abst. Jour. Soc. Chem. Ind., Vol. XXXIII, 1914, p. 1014.
[6] British Patent, No. 3,923, March 27th, 1885. Abst. Jour. Soc. Chem. Ind., Vol. V, 1886, p. 172.
[7] German Patent, No. 18,303, June 1st, 1881. Fischer's Jahresbericht, Vol. XXVIII, 1882, p. 120.
[8] British Patent, No. 10,491, Sept. 4th, 1885. Abst. Jour. Soc. Chem. Ind., Vol. IV, 1885, p. 746.
[9] German Patent, No. 28,989, 1884.
[10] German Patent, No. 32,006, 1885. Abst. Fischer's Jahresbericht, Vol. XXXII, 1886, p. 158.
[11] British Patent, No. 1464, 1884. Abst. Jour. Soc. Chem. Ind., Vol. III, 1884, p. 261.

Winkler[1] states that in the preparation of cobalt and nickel castings, a high temperature is necessary, refractory crucibles should be used, carbon and silicon should not come in contact with the molten metal, and that the metal must be protected from the atmospheric oxygen during casting.

Fleitmann[2] proposed a method to remove the brittleness in cobalt and nickel metal by adding magnesium before pouring.

Fink[3] outlines a process for the treatment of ores from Cobalt, Canada. The ore is ground to 40-mesh and mixed with fluxes to reduce the metals and make a slag. The charge is heated in a furnace at a temperature of 1,200 to 1,500° C. under reduced pressure for several hours. The arsenic is volatilized and condensed as metallic arsenic; and metallic silver, cobalt, and nickel speiss and slag are also formed. The products are further treated by known methods.

Levat, in a paper read before the French Association for the Advancement of Sciences, Sept. 29, 1887, gave the results of an investigation of the nickel, cobalt, and chromium ore deposits of New Caledonia.

Heard[4] gave a summary of the possibilities of the New Caledonia deposits.

2.—The Production of Smalt

Smalt is a potash aluminum cobalt silicate. It is prepared by melting together silica, potassium carbonate, and roasted cobalt ore. It was used extensively a number of years ago to produce blue-coloured glass and enamels. Owing to the difficulty in obtaining the same quality of colour intensity with different cobalt ores, the manufacture of smalt has gradually been abandoned. Cobalt oxides of uniform composition can be readily obtained, and these are used at present to prepare the different colours.

The production of smalt was formerly carried on chiefly in Saxony, and it is interesting to note that the Chinese have also prepared smalt for a number of years.

The first record of cobalt being used in Europe to colour glass dates from about 1540. It is stated a Nuremberg glass-maker was the first to try melting cobalt ore, termed " Kobold," with glass, and he obtained, much to his astonishment, a beautiful blue-coloured product.

A brief account of the operations as conducted in Saxony and China is given in the following paragraphs.

The Method Employed in Saxony to Prepare Smalt

The mineral was crushed, sorted and washed. This product, called " Schlich," was roasted in a rotary furnace, any arsenic evolved during the roasting being collected. During the roasting the cobalt was converted into oxide. The roasted product was finely ground and screened through a silk sieve, the powder being known as " zaffer " or "zaffler."

---

[1] Chem. News, Vol. XXXV, 1877, p. 166.
[2] Chem. News, Vol. XL, 1879, p. 67.
[3] United States Patent, No. 1,013 931, Jan. 9th, 1912.
[4] Eng. Min. Jour., Vol. XLVI, 1888, p. 103

To prepare smalt the roasted ore was mixed with sand and potassium carbonate and melted in a crucible. Crushed quartz was used for sand. Before being crushed, it was heated to a red heat in a lime kiln, crushed in stamp-batteries, washed to remove any light impurities, dried, and again heated to redness. The cobalt, being more readily oxidized than the nickel, passed into the slag, the nickel, combining with any arsenic, forming a speiss which settled to the bottom of the crucible. Any iron present was oxidized to ferric oxide, which is less injurious than ferrous oxide. The speiss was withdrawn either through an opening in the bottom of the crucible or carefully removed with a ladle. The purer the materials employed in the preparation, the more beautiful the product. The blue glass or slag containing the cobalt was poured into water, dried, and finely ground. The ground glass or smalt was mixed with water and allowed to stand for half an hour, during which time the coarse particles, known as " Streublau," settled. The coarse particles were afterwards reground.

The turbid water from the first washing was decanted into a second tank in which the pigment proper, termed couleur, settled to the bottom.

After twenty-four hours the turbid water was decanted into a third tank in which it remained until it was clear, when the finest and palest glass powder, " Aeschel," settled. The pigment, and also the aeschel, was next washed two or three times, the wash waters being filtered. The pigments in the various settling tanks were dried, screened, and packed into barrels.

About three-fifths of the glass taken from the pots was recovered. The presence of oxides other than those of cobalt and potash, even in small quantities, exerts a marked influence on the colour of the smalt. Barium produces an indigo tinge; sodium, calcium, and magnesium produce a reddish shade; iron, a blackish green, very objectionable in the brighter-coloured smalts; manganese, violet; nickel, violet, but less intense; copper, zinc, bismuth, and antimony, dull shades.

The smalt is classified according to its degree of fineness into coarse blue (Streublau) pigment, and Aeschel, the first size being denoted in the trade by H, the second by C, and the finest by E.

Respecting the intensity of the colour, each sort is distinguished as fine, middle, and ordinary by the letters F, M, and O. In the first class, colours of varying degrees of intensity are denoted by the letters, F, FF, FFF, FFFF, expressing two-fold, three-fold, and four-fold respectively. Qualities poorer in cobalt than the OC quality are distinguished by the use of indices, e.g., OC^2 (i.e., containing half the cobalt in the OC quality).

Smalts which contain more cobalt than the F quality are distinguished by doubling the latter F.

More intense than the last, FFFF, which is termed " Azure," or " King's Blue," do not occur.

The following list[1] gives the different brands of smalt:—

H ....Ordinary smalt.
E ....Aeschel.
B ....Bohemian smalt.
CF ....Ground pigment.

---

[1] Grünwald, Raw Materials of the Enamel Industry, 1914, p. 146.

FC .... Fine pigment.
FCB .... Fine Bohemian pigment.
FE .... Finer Aeschel.
MC .... Average.
MCB .... Average Bohemian.
ME .... Average Aeschel.
OC .... Ordinary pigment.
OCB .... Ordinary Bohemian pigment.
OE .... Ordinary Aeschel.

The different colours are spoken of as azure blue, smalt blue, zaffer blue, Saxon blue, enamel blue, cobalt blue, etc.

The following are analyses of some varieties of smalt:

Percentage Composition of Some Typical Smalts

Component	Norwegian Smalt		German Smalt			French Smalt				
	Dark	Dark Aeschel	Pale	Coarse Blue	Pigment C.	I	II	III	IV	V
Silica	70.86	66.20	72.11	72.21	70.11	75	70	65	63	60
Cobalt oxide...........	6.49	6.75	1.95	20.54	21.58	20	26	30	30	30
Potash	21.41	16.31	1.80	6.75	7.20	2	3	5	7	10
Alumina...............	0.43	8.64	20.04	0.22	0.11	....	....	....	....	....
Arsenious acid..........	......	......	......	......	......	1	1	1	1	1

There is also a French method of smalt preparation which consists in melting together cobalt oxide with quartz and potash. In this way a first-class product of desired intensity can be prepared, although the process is correspondingly more expensive.

The quartz is heated to redness and ground as in earlier methods, only instead of cobalt speiss pure cobalt oxide is here employed. The smalt so obtained is very pure, and more durable than the ordinary product.

The Chinese Method of Manufacturing Smalt[1]

In the Chinese glass works at Canton, the so-called lam-o-li-shek, i.e., "stone for blue glass," is employed for the production of a blue colour in glass and porcelain. It appears as if the Chinese are unaware how to produce colourless glass. They purchase glass fragments from Europe and America, which they classify according to colour and quality, and melt them in pots, 67 cm. at the top. One or two of these pots are placed in a rectangular furnace of primitive construction, which is heated with anthracite, for which purpose from 150 to 200 kg. of anthracite per pot are required. The blowpipe is short and wide, while the moulds are made of clay and dust. It is astonishing how the Chinese, by means of their small apparatus, are able to separate the cobalt from the iron, manganese and nickel, even when the cobalt content in the ore does not exceed 2 to 4 per cent.

[1] Bowler, Chemical News, Vol. LVIII, 1888, p. 100.

The crude cobalt mineral is first carefully washed, and every piece is scrubbed with a brush, in order to remove the adhering clay which contains iron. The ore is next dried and pulverized, then afterwards ground in a hand mill with water. The whole mixture is then conveyed to a vessel, in which it is vigorously stirred for several hours, after which it is allowed to settle over night. On the following day the water is decanted, taking with it the upper layer of the settled powder. The latter consists of the lighter earthy substances, while the residual mass is the oxides of iron, manganese, nickel and cobalt. This mass is removed, mixed with a small quantity of borax, and placed in one of the above-mentioned pots containing the glass. The melting is now carried out, and during the first twelve hours the fused mass acquires a dirty greenish-black colour. By degrees this mixed colour changes to a bright bluish-violet similar to amethyst. It appears as if the iron and nickel are completely reduced, for after 36 to 40 hours' heating scarcely any of the iron or manganese colour can be detected in the mixture.

The lowest portion of the fused glass in the pot is rejected. This part probably contains the impurities, such as iron, nickel, manganese, etc.

For the purpose of porcelain painting, the Chinese frit, the same mineral with feldspar, kaolin, and much borax. This frit is ground to fine powder, and employed for painting on the biscuit. The Canton process is as follows:

The manipulator takes the burnt biscuit and covers this with a glaze consisting of borax, feldspar, and clay, which, when sufficiently dry, he paints upon, and in one single operation burns in both the enamel and colour.

An analysis of the mineral is given as follows: Iron oxide 35.0, Manganese oxide 13.1, Nickel oxide, 3.5, Cobalt oxide, 3.5, gangue 46.0 per cent.

The gangue consisted for the most part of silica and aluminium silicate.

Preparation of Metallic Cobalt

Metallic cobalt may be obtained from the oxide or oxalate by one of the following methods. Cobalt oxide is obtained by heating precipitated cobalt hydroxide or oxalate.

Reduction of the oxide in a carbon crucible or by the addition of carbon or starch.[1]

Reduction of the oxide[2] or oxalate[3] in a stream of hydrogen or hydrocarbons,[4] the reduction being complete at 500 to 600° C.

Reduction of the oxide by carbon monoxide.[5]

Reduction of the oxide by ammonium chloride.[6]

Reduction of cobalt chloride in a current of hydrogen.[7]

Reduction of the oxide by aluminium.[8] Goldschmidt process.

Precipitation from cobalt solutions by metallic magnesium.[9]

[1] Berthier, Annales de Chemie et Physique, Vol. 25, 1824, p. 98. Winkler, Jour. prakt. Chem., Vol. XCI, 1864, p. 213. Kalmus, Preparation of Metallic Cobalt by Reduction of the Oxide: Department of Mines, Canada, Bulletin No. 259, 1913, p. 4. Moissan, Comp. Rend., Vol. 116, 1893, pp. 349-351.

[2] Muller, Annalen der Physik und Chemie, Vol. CXXXVI, 1869, p. 51. Moissan, Annales de Chem. et Physique, 1886, Vol. V, sec. 21, p. 199. Glasser, Zeitschr. anorg. Chem., Vol. XXXVI, 1903, p. 19. Kalmus, Bulletin 259, p. 11.

[3] Wolff, Zeitschr. anal. Chem., Vol. XVIII, 1879, p. 38. Berzelius, Gmelin Kraut, Handbuch der anorganischen Chemie, 1909, Band. 5, 1, p. 194. Brunner, Idem, p. 194.

[4] Martin, German Patent, No. 18303, June 1st, 1881. Fischer's Jahresbericht, 1882, Vol. 28, p. 120.

[5] Mond, Hirtz, and Copaux. Note on a volatile compound of cobalt with carbon monoxide: Chem. News, Vol. XCVIII, 1908, p. 165. Mond Nickel Co., Hirtz, and Copaux, British Patent, No. 13,207, 1908: a patent covering the manufacture of cobalt carbonyl or carbonyls by heating cobalt or material containing cobalt in carbon monoxide or gas containing the latter. Kalmus, Bulletin 259, p. 25.

[6] Rose, Gmelin Kraut, Handbuch der anorganischen Chemie, Band V. 1, 1909, p. 192.

[7] Peligot, Compt. Rend., Vol. XIX, 1844, p. 670. Baumhauer, Zeitschr. anal. Chem., Vol. X, 1871, p. 217. Schneider, Annalen der Physik und Chem., Vol. CI, 1857, p. 387. Wagner, Jahresbericht, 1857, p. 226.

[8] Kalmus, Bulletin 259, p. 32.

[9] Siemens, Zeitschr. anorg. Chem., Vol. XLI, 1904, p. 249.

Distillation of cobalt amalgams.[1]

Electrolytic methods.[2]

Fink[3] proposes the following treatment for the production of metallic cobalt. Powdered smaltite is mixed in the proper proportion with powdered lime and calcium carbide and heated in a vacuum for 1 to 2 hours, at a temperature of 1500° C. The reaction of lime and calcium carbide yields metallic calcium, which in turn reacts with the cobalt arsenide, forming calcium arsenide and metallic cobalt.

Precipitation of cobalt from solutions by zinc.[4]

Additional References

Copaux, Pure Cobalt and Nickel, Preparation and Properties; Revue générale de Chimie pure et appliquée, Jaubert, Paris, 1906, p. 156. Comp. Rend., Vol. 140, 1905, pp. 657-659.

Winkler, The Atomic Weight of Nickel and Cobalt: Zeitschr. anorg. Chemie, Vol. VIII, 1895, p. 29.

Hamilton, Aluminium Precipitation at Nipissing Mine, Cobalt, Canada: Eng. and Min. Jour., Vol. XCV, 1913, pp. 935-939.

Kirkpatrick, Aluminium Precipitation at Deloro, Canada: Eng. and Min. Jour., Vol. 95, 1913, p. 1277.

Larson and Helme, Electrolytic Recovery of Cobalt and Zinc from the End-lyes of Copper Extraction: Chem. Abstracts, 1913, p. 3085.

Megraw, Cyaniding a Furnace Product (a roasted cobalt-nickel speiss). Eng. and Min. Jour., Vol. XCVIII, 1914, p. 147.

Reid, Milling Practice in Cobalt: Trans. Can. Min. Inst., Vol. 14, 1914, pp. 50-63.

Denny, Desulphurizing Silver Ores at Cobalt: Min. Sci. Press, Vol. CVII, 1913, pp. 484-488.

Kleinschmidt, Summary of the Treatment of Cobalt Ores: Berg-u. huttenm. Zeitung, Vol. XXVI, 1867, pp. 45-46, 57-59, 130-131, 147-149, 162-164.

McCay, Contribution to the Study of Cobalt, Nickel, and Iron Sulphides, Freiberg, 46 pp. (pamphlet), 1883.

Report and Appendix of Royal Ontario Nickel Commission, 1917.

Development of the Metallurgy of the Ontario Silver-Cobalt Ores

In reviewing the development of the metallurgy of the silver-cobalt ores of Ontario, it is necessary to make three divisions, viz., the progress in ore-dressing methods, the introduction of new processes in the extraction of the silver, and the improvements in the treatment of the ores for the cobalt. At the same time, it cannot be too strongly emphasized that co-operation and publicity of results have contributed no small part to the success achieved in the treatment of the Cobalt ores.

The importance of the Cobalt deposits may best be realized from the following figures. In 1903 when the silver deposits of Cobalt were discovered, Canada produced silver valued at $1,709,643; while in 1913 the production was valued at $19,040,924. During the same period the value of the annual production of cobalt rose from practically nothing to $500,000. Besides the value of cobalt, appreciable returns are realized from the arsenic and nickel.

Previous to the discovery of the deposits at Cobalt, most of the world's supply of cobalt ore was shipped from New Caledonia to Europe for treatment. In fact,

---

[1] Moissan, Compt. Rend., Vol. LXXXIII, 1879, p. 180; Bulletin Soc. Chem., Pt. II, Vol. XXXI, 1879; p. 149; Annalen de Chem. et. Physique, Vol. V, 1880, p. 21, p. 199. Guertler, Metallographie, Vol. I, 1912, p. 513.

[2] Kalmus, Electro-plating with Cobalt, Bulletin No. 334, Department of Mines, Canada, 1915. Becquerel, Comp. Rend., Vol. LVIII, 1862, pp. 18-20.

[3] United States Patent, No. 1,119,588, Mineral Industry, Vol. XXIII, 1914, p. 548.

[4] Careis, Berg-u. huttenm. Zeitung, Vol. XL, No. 23, 1881, pp. 215, 216.

until 1908 Europe practically controlled the world's production. In 1908 cobalt products from Canadian and American smelters were placed on the market, and this caused the price of cobalt oxide to fall from $2.00 to $1.00 a pound. Cobalt oxide is at present (1917) selling for $1.50 per pound.

Progress in Ore Dressing

The progress in ore dressing at Cobalt has been brought about by the diminution in the grade of the ore and the attempt to obtain increased recoveries. At first the high-grade ore (2,000 oz. or more) was readily hand-picked from the low-grade and shipped as such. In order to keep up the grade of the shipments, concentration methods were introduced extensively during 1907-1908. For the coarser crushing it is customary to use one of the following combinations: stamps crushing to approximately 20-mesh; crushing in stamps to 0.25-inch, followed by sizing and concentrating, the fine tailing going to waste, the coarse being reground in pebble mills; and crushing in rolls or in ball mills to about 0.25-inch, concentrating, and regrinding the tailing in pebble mills. In 1914 about 75 per cent. of the ore milled was crushed by stamps to 14 to 20-mesh.

Classification is used extensively to separate the sand from slime. In the concentrating mills hindered-settling types of classifiers are employed, while in the cyaniding plants the Dorr classifier is common. For sand concentration the Wilfley, James, and Deister tables are mostly used, while for treating slime the James and Deister tables are preferred, although canvas tables were employed to some extent. The sand tailing averages 3 to 4 ounces and the slime tailing 6 to 8 ounces of silver per ton, but it is not possible to recover the silver from these products by the ordinary gravity concentration methods. The average ratio of concentration in the Cobalt mills is 50 to 1.

The recovery of the silver from the tailing dumps and low-grade ores has been one of the problems that has confronted the metallurgists at Cobalt for some time. At present there are about 2,500,000 tons of tailing, averaging 4 to 6 ounces of silver. To treat these accumulated residues and the low-grade ores, flotation methods seem most suitable. In October, 1915, the first experimental flotation plant was erected by the Buffalo Mines, Limited. This plant consisted of a two-compartment, standard length Callow rougher cell, and a one-half size Callow cleaner cell. The results obtained by flotation were so encouraging that a flotation plant to handle 600 tons daily was erected and put into operation in September, 1916. At present the Buffalo Mines, McKinley-Darragh-Savage Mines, Nipissing Mines, Coniagas Mines, Dominion Reduction Co., Northern Customs Concentrators, and National Mines (King Edward) employ the Callow system of flotation.

To prepare the tailings and low-grade ores for flotation, fine grinding to 100-mesh is necessary. Pebble mills are commonly used for this purpose. The oil used consists of a mixture of 15 per cent. pine oil, 75 per cent. coal tar creosote, and 10 per cent. coal tar. Further experiments have shown that an oil obtained from the distillation of hard wood in charcoal plants is well suited for flotation. The results obtained by flotation are about as follows: feed 6 to 10 oz. of silver, concentrate 250 to 1,000 oz., tailing 0.8 to 2.5 oz., and a recovery of 80 to 90 per cent.

6 B.M. (iii)

The recovery of the silver from the flotation concentrate has been accompanied with difficulties, and at present most of the concentrate is shipped to the smelters in the United States. Under present conditions the freight and treatment charges on the flotation concentrate amounts to about 20 per cent. of the value of the product. The flotation concentrate produced at the Buffalo Mines was treated at Cobalt. The method employed is to give the concentrate a roast with salt, a leach with hydrochloric acid, and a treatment with salt solution, the silver being finally recovered by precipitation on scrap copper, and the copper by passing the solution over scrap iron. The above treatment has been discontinued by the Buffalo Mines company, but is still being practised by the Dominion Reduction company.

Progress in the Metallurgy of Silver

In the early operation of the Cobalt camp, most of the silver was recovered by smelting methods, the ore shipped averaging 2,000 ounces or more per ton. The presence of large quantities of arsenic caused trouble in the smelting and refining processes, and it was necessary for the smelters to levy a penalty on excessive quantities. There was also a high treatment charge because of the difficulty in smelting the high-grade silver ores. The freight rate on the high-grade ore was also heavier.

In order to obtain a better return from the silver contents of the high-grade ore, experiments were undertaken to devise a process for treating the high-grade ores at Cobalt and obtaining the silver as bullion. The Nova Scotia mill, now the Dominion Reduction, was the first to recover bullion by amalgamation and cyanidation. This mill treated the run-of-mine ore, which included the high-grade, and produced bullion and a residue assaying about 150 ounces of silver per ton. The Nipissing improved on the process by substituting a tube mill for the amalgamating pan. The combination process of amalgamation and cyanidation is used at three mills, viz., the Nipissing and Buffalo on high-grade ore and concentrates, and the Dominion Reduction on concentrates only. The process is usually carried out as follows: With 2,500 oz. ore, a charge would be 6,500 pounds of 20-mesh ore, 8,500 pounds of mercury, 3,800 pounds of 5 per cent. cyanide solution, and 6 tons of pebbles. After 9 to 10 hours' treatment the pulp will all pass a 200-mesh screen, and contain about 50 oz. of silver per ton. It is given a further agitation for 36 hours in a 0.75 per cent. cyanide solution when the silver is reduced to 30 ounces.

The treatment of low-grade ores at Cobalt has for a number of years presented a difficult but interesting problem. It is possible to treat the low-grade ores by concentration methods, by a combination of concentration and cyaniding, and by flotation. It appeared as if flotation in combination with concentration or cyanidation was the most suitable, but recent developments tend in some cases to favor concentration and cyanidation. Very little cobalt and nickel are recovered in the flotation concentrate.

Amalgamation was not practicable on low-grade ores, so attention was turned toward cyanidation. After extensive experiments it was found that it was not possible to treat the low-grade complex ores by the ordinary cyanide process, because the solutions quickly became foul, thereby diminishing the dissolving power; besides, the cyanide consumption was excessive. To overcome these difficulties a process known as the "Wet Desulphurizing Process" was devised by the Nipissing Mining Company.

Briefly, this process consists in giving the ground ore a treatment with alkali and aluminium. The preliminary treatment is given in tube mills, and the final treatment in tanks. In the preliminary treatment the ore is ground in an 0.25 per cent. caustic soda solution with an addition of 5 pounds of lime per ton of ore. The lime is added to hasten settling. In the process the refractory minerals, especially pyrargyrite and proustite, are decomposed, the sulphur, arsenic, and antimony being reduced to the elemental or metallic state. The reduced sulphur, arsenic, and antimony have practically no action on cyanide. The reactions involved may be expressed as follows:

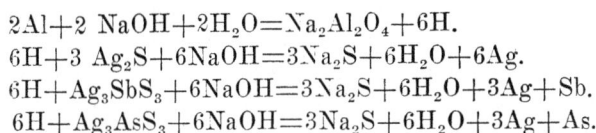

$$2Al + 2\ NaOH + 2H_2O = Na_2Al_2O_4 + 6H.$$
$$6H + 3\ Ag_2S + 6NaOH = 3Na_2S + 6H_2O + 6Ag.$$
$$6H + Ag_3SbS_3 + 6NaOH = 3Na_2S + 6H_2O + 3Ag + Sb.$$
$$6H + Ag_3AsS_3 + 6NaOH = 3Na_2S + 6H_2O + 3Ag + As.$$

With fine grinding and after treating the ore by the wet desulphurizing process, the silver may be readily extracted in 48 hours by treatment with a 0.25 per cent. cyanide solution. By the desulphurizing treatment a saving of one to four ounces of silver is obtained at an additional cost of 54 cents per ton of ore. On a 20-ounce feed a saving of four ounces means an increased extraction of 20 per cent.

The precipitation of the silver from the cyanide solutions by zinc presented a further difficulty, but the substitution of aluminium dust overcame this. The advantages of the aluminium dust in precipitation may be stated as follows: a high-grade precipitate is obtained, there is little fouling of the cyanide solutions, and during the precipitation an amount of cyanide proportional to the silver precipitated is regenerated. One part of aluminium dust will precipitate about three parts of silver. Aluminium dust was used until 1916, but owing to the difficulty of obtaining it and also to the increased cost, other precipitants are being tried. At present at the Nipissing mill sodium sulphide is used to precipitate the silver, and the sulphide precipitate is desulphurized by treatment with caustic soda and aluminium, metallic silver being produced.

Aluminium dust precipitation was first used by the Deloro Smelting and Refining Company, then by the O'Brien mine, and later by the Nipissing and Buffalo and other mines at Cobalt.

Progress in the Metallurgy of Cobalt and Nickel

Before the discovery of the large silver-cobalt-nickel deposits at Cobalt, very little was known in Canada of the properties and metallurgy of the metal cobalt and its compounds, except that cobalt silicate possessed a beautiful colour and was used in the ceramic industries. Cobalt was considered one of the rarer metals.

Cobalt ores had been treated in Europe for centuries, but the treatment was conducted on a small scale, and the methods were what might be called large laboratory methods. There were about 25 plants operating in Europe, and these were producing about one-half the quantity of cobalt that is now being turned out annually by the three Canadian smelters. The method formerly employed was to dissolve the ore, matte or speiss in either hydrochloric or sulphuric acid, and remove the impurities by chemical methods. The copper was often removed by hydrogen sulphide gas, and after eliminating any gas in solution by boiling, the iron, cobalt, and nickel were precipitated. The high price of cobalt oxide enabled the operators to work on a small scale, even though the grade of the ore would not average over 3 per cent. cobalt.

After the discovery of the cobalt deposits in Canada, it soon became evident that they were of sufficient extent to justify the attempt to establish a cobalt smelting industry in Canada. It was soon seen that new processes would have to be devised, or the old methods improved. The cost of acids and chemicals, the operating difficulties, and the higher cost of labour made the old processes prohibitive in Canada, and at the same time it was not possible to operate them on a considerable scale. The larger proportion of cobalt and the presence of large quantities of arsenic in the Canadian ores, also made it necessary to modify the older processes.

In the processes in use in Canada at present, the gangue minerals, and most of the arsenic and iron are removed in blast furnaces. The products of the blast furnace are: metallic silver, an argentiferous speiss containing cobalt, nickel, and iron as arsenides, also slag and flue dust. The large quantities of silver and arsenic in the cobalt-nickel ores are a source of revenue to the smelters. The argentiferous speiss made in the blast furnace is roasted to about 10 per cent. arsenic, given a chloridizing roast, and the product cyanided. The successful cyanidation of such a complex furnace product was a new departure in cyaniding. The residue is sulphated with sulphuric acid, and the iron is rendered insoluble by heating. In this treatment with acid the cobalt and nickel dissolve, as well as some of the iron and arsenic. The iron and arsenic must be removed before the precipitation of the cobalt and nickel, with solutions of bleaching powder. The grade of cobalt oxide produced by the Canadian smelters is considerably better than that produced by the European refineries. The production of a higher grade oxide at a lower price was necessary to compete with and secure the closely controlled trade of Europe. Both these difficulties were finally overcome by the improvements in and the efficiency of the Canadian processes.

In summarizing, it may be stated that the progress in the metallurgy of cobalt and nickel has not been owing so much to the introduction of new methods as to the development of the old ones on a larger and more efficient scale.

The general progress in the development of the metallurgy and ore dressing of the silver-cobalt-nickel ores has been due to the grade and value of the ore, and to the co-operation of the operators. It would be impossible to mention all those who have contributed to the success of the Cobalt camp, but special mention should be made of the work of Reid and Moffat in ore dressing; of Denny, Fairlie, Jones, and Clevenger in the metallurgy of silver; and of Kirkpatrick and Peek in the metallurgy of cobalt and nickel.

The wet desulphurizing process used in connection with the low-grade ores, and the introduction of the use of sodium sulphide to precipitate the silver, were developed by J. J. Denny, metallurgist of the Nipissing Mining Company. Prof. S. F. Kirkpatrick was the first to introduce the use of aluminium dust for the precipitation of silver, but his chief work has been in the development of the metallurgy of cobalt and nickel. Under his direction the Deloro Smelting and Refining Company has become the largest smelter treating cobalt ores and also, due mostly to his initiative, a plant has been erected to smelt the complex cobalt ores of Missouri. Difficulties have constantly arisen in the treatment of ores of cobalt, but these have always been solved. The excellent work of Prof. Kirkpatrick in the development of the metallurgy of cobalt-nickel ores has been recognized by the profession, since in 1917 he was awarded the McCharles medal of the University of Toronto.

CHAPTER III

THE CHEMISTRY OF COBALT

The word cobalt is synonymous with " kobold," meaning goblin, which was a term given by the early miners to those ores which did not yield metal on smelting. It is stated[1] that an early form of the word cobalt appears in the writings of Basilius Valentinus about the end of the 15th century. Berthelot[2] states that the word is of Graeco-Egyptian origin. In Hoover's translation of Agricola's De Re Metallica, mention is made of the word cobalt as being from the Greek, cobalos.

The use of cobalt compounds for colouring glass was known to the ancients and, since 1600, cobalt minerals have been used for the preparation of smalt. In 1735 Brandt prepared some metallic cobalt by reduction from the ore.

Polished cobalt metal is silvery white in colour, but when reduced from the oxide, it is in the form of gray powder. The specific gravity varies from 8.79 on an unannealed sample to 8.92 on a swaged sample. The melting point of cobalt is given as 1478°C., and the tensile strength at about 34,400 pounds per square inch.[3] Metallic cobalt is magnetic. The atomic weight is 58.97.

Cobalt is soluble in dilute acids. The metal forms three oxides: cobaltous oxide (CoO) greenish gray; cobaltous cobaltic oxide (Co_3O_4) black; and cobaltic oxide (Co_2O_3) brownish. Cobaltous oxide is obtained from Co_3O_4 by heating at a high temperature. The usual method of preparing the oxides is to calcine, at a red heat, the hydroxide obtained by precipitation in one of the processes mentioned under the Metallurgy of Cobalt.

Cobalt forms with acids two compounds, cobaltous and cobaltic. Cobaltous compounds are pink in the crystallized state or in aqueous solutions, but yellow or green in the anhydrous condition, and blue when in aqueous solutions in the presence of hydrochloric acid.

By dissolving any of the three oxides in acids, salts derived from cobaltous oxide are always obtained, containing bivalent cobalt:

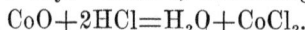

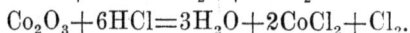

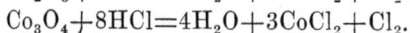

$$CoO+2HCl=H_2O+CoCl_2.$$
$$Co_2O_3+6HCl=3H_2O+2CoCl_2+Cl_2.$$
$$Co_3O_4+8HCl=4H_2O+3CoCl_2+Cl_2.$$

Simple cobaltic salts are unknown, but many complex compounds exist with trivalent cobalt, as, for example, potassium cobaltinitrite, potassium cobalticyanide, and numerous cobalti-ammonia derivatives.

Reactions of Cobalt Salts [4]

Potassium or sodium hydroxide precipitates in the cold a blue basic salt:

$$CoCl_2+KOH=KCl+Co(OH)Cl,$$

which on warming is further decomposed by hydroxyl ions, forming pink cobaltous hydroxide:

$$Co(OH)Cl+KOH=KCl+Co(OH)_2.$$

[1] Gmelin Kraut, Handbuch der anorganischen Chemie, Band V, 1, 1909. p. 190.
[2] Idem, p. 190.
[3] Kalmus, The Physical Properties of Cobalt, Bulletin 309, Department of Mines. Ottawa, Canada, 1914.
[4] A number of the following reactions are taken from Treadwell Hall, Analytical Chemistry, Vol. I, fourth edition, 1916.

In the case of a moderately concentrated solution of the alkali the precipitate of pink cobaltous hydroxide is often produced in the cold, sometimes only after standing for some time. The rapidity of the reaction depends entirely upon the concentration of the alkali.

Cobaltous hydroxide gradually turns brown in contact with the air, forming cobaltic hydroxide:

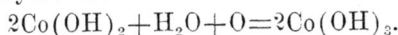

$$2Co(OH)_2 + H_2O + O = 2Co(OH)_3.$$

In this respect cobalt behaves similarly to iron and manganese, but differs from nickel, for the hydroxide of the latter is not oxidized by atmospheric oxygen.

On adding chlorine, bromine, hypochlorites, hydrogen peroxide, etc., to an alkaline solution containing cobaltous hydroxide, cobaltic hydroxide is immediately formed, as with nickel and manganese:

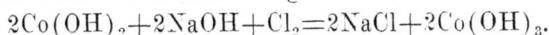

$$2Co(OH)_2 + 2NaOH + Cl_2 = 2NaCl + 2Co(OH)_3.$$
$$2Co(OH)_2 + H_2O + NaOCl = NaCl + 2Co(OH)_3.$$

From ammoniacal cobalt solutions the above oxidizing agents cause no precipitation, but merely a red colouration; the addition of potassium hydroxide then causes no precipitation, whereas in the case of nickel, a precipitate is formed.

Cobaltous hydroxide—$Co(OH)_2$, behaves under some conditions as a weak acid, for on adding to a cobaltous solution a very concentrated solution of KOH or NaOH the precipitate at first produced dissolves with a blue colour similar to that formed with copper compounds. By the addition of Rochelle salts, $KNaC_4H_4O_6$, to this blue cobalt solution the colour either disappears almost entirely or becomes a pale pink, while the similarly treated copper solution becomes more intensely blue. By the addition of potassium cyanide to the blue cobalt solution it becomes yellow, and in contact with air turns intensely brown. A copper solution would be decolourized by the addition of potassium cyanide.

By pouring a little cobalt solution (or adding a little solid cobalt carbonate) into a concentrated solution of caustic soda or potash, to which a little glycerol has been added, a blue solution is formed (the colour being intensified by warming), which after standing some time in the air, or immediately on the addition of hydrogen peroxide, becomes a beautiful green.

Ammonia precipitates, in the absence of ammonium salts, a blue basic salt, soluble, however, in excess of ammonium chloride. Ammonia, therefore, produces no precipitate in solutions which contain sufficient ammonium chloride. The dirty yellow ammoniacal solution is little by little turned reddish on exposure to the air, owing to the formation of stable cobalti-ammonia derivatives:

$$Co(OH)_2 + 2NH_4Cl + 2NH_3 = Co(NH_3)_4Cl_2 + 2H_2O.$$

Alkali carbonates produce a reddish precipitate of basic salt of varying composition.

Ammonium carbonate also precipitates a reddish basic salt, soluble, however, in excess.

Barium carbonate does not precipitate cobalt in the cold and out of contact with air, but on exposure to the air cobaltic hydroxide is gradually thrown down. The precipitation takes place much more quickly on the addition of hypochlorites or hydrogen peroxide:

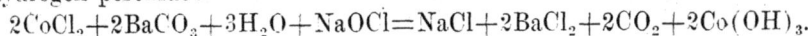

$$2CoCl_2 + 2BaCO_3 + 3H_2O + NaOCl = NaCl + 2BaCl_2 + 2CO_2 + 2Co(OH)_3.$$

If the solution is heated to boiling, all of the cobalt is precipitated as a basic salt, even out of contact with the air.

Hydrogen sulphide produces no precipitate in solutions containing mineral acids. In neutral solutions containing an alkali acetate, all of the cobalt is precipitated as black sulphide.

Ammonium sulphide precipitates a black sulphide,

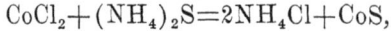

$$CoCl_2 + (NH_4)_2S = 2NH_4Cl + CoS,$$

insoluble in ammonium sulphide, acetic acid, and very dilute hydrochloric acid; soluble in concentrated nitric acid and aqua regia, with the separation of sulphur:

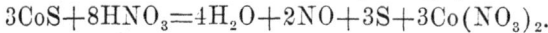

$$3CoS + 8HNO_3 = 4H_2O + 2NO + 3S + 3Co(NO_3)_2.$$

By continued action of strong nitric acid all the sulphur goes into solution as sulphuric acid.

Potassium cyanide produces in neutral solutions a reddish brown precipitate, soluble in excess of potassium cyanide in the cold, forming brown potassium cobaltocyanide:

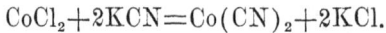

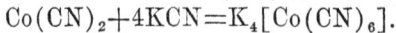

$$CoCl_2 + 2KCN = Co(CN)_2 + 2KCl.$$
$$Co(CN)_2 + 4KCN = K_4[Co(CN)_6].$$

On warming the brown solution for some time it becomes bright yellow and gives an alkaline reaction. It now contains potassium cobalticyanide, of analogous composition to potassium ferricyanide.

The formation of the cobaltic salt takes place in the presence of atmospheric oxygen:

$$2K_4Co(CN)_6 + O + H_2O = 2KOH + 2K_3Co(CN)_6.$$

The reaction takes place more quickly in the presence of chlorine, bromine, hypochlorites, etc.:

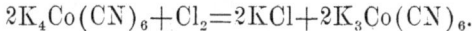

$$2K_4Co(CN)_6 + Cl_2 = 2KCl + 2K_3Co(CN)_6.$$

An excess of chlorine, bromine, etc., does not decompose the cobaltic salt; in this particular it differs from nickel.

The cobalticyanide anion is much more stable than the cobaltocyanide anion. By adding hydrochloric acid to the brown solution of potassium cobaltocyanide, hydrogen cyanide (prussic acid) will be set free and yellow cobaltous cyanide formed,

$$K_4Co(CN)_6 + 4HCl = 4HCN + 4KCl + Co(CN)_2,$$

while potassium cobalticyanide is not decomposed by hydrochloric acid.

Potassium cobalticyanide forms, with most of the heavy metals, difficulty soluble or insoluble salts possessing characteristic colours. Thus, it produces with cobaltous salts pink cobaltous cobalticyanide:

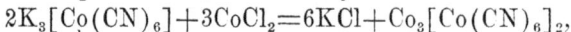

$$2K_3[Co(CN)_6] + 3CoCl_2 = 6KCl + Co_3[Co(CN)_6]_2,$$

and with nickel salts greenish nickel cobalticyanide. If, therefore, a cobalt solution contains nickel it forms, when treated with sufficient potassium cyanide to redissolve the cobalt precipitate, boiled, and acidified with hydrochloric acid, a greenish precipitate of nickelous cobalticyanide:

$$2K_3[Co(CN)_6] + 3K_2[Ni(CN)_4] + 12HCl = 12HCN + 12KCl + Ni_3[Co(CN)_6]_2.$$

Potassium nitrite produces in concentrated solutions of cobalt salts, in the presence of acetic acid, an immediate precipitation of yellow crystalline potassium cobaltic nitrite. If the solution is dilute, the precipitate appears only after standing for some time, but more quickly on rubbing the sides of the beaker.

The reaction takes place in the following stages:

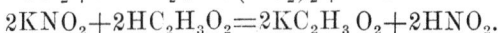

$$CoCl_2 + 2KNO_2 \rightleftharpoons Co(NO_2)_2 + 2KCl.$$
$$2KNO_2 + 2HC_2H_3O_2 = 2KC_2H_3O_2 + 2HNO_2.$$

The free nitrous acid oxidizes the cobaltous nitrite to cobaltic nitrite,

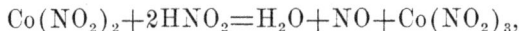

$$Co(NO_2)_2 + 2HNO_2 = H_2O + NO + Co(NO_2)_3,$$

which now combines with more potassium nitrite:

$$Co(NO_2)_3 + 3KNO_2 = K_3Co(NO_2)_6.$$

This reaction offers an excellent means of detecting the presence of cobalt in nickel salts.

Potassium nitrite produces in dilute nickel solutions no precipitate. In very concentrated solutions a brownish-red precipitate of $Ni(NO)_2.4KNO_2$ is thrown down; in the presence of alkaline earth salts a yellow crystalline precipitate is formed; e.g., $Ni(NO_2)_2.Ba(NO_2)_2.2KNO_2$, which is very difficultly soluble in cold water, but readily soluble in boiling water, with a green colour.

Ammonium thiocyanate (Vogel's reaction): If a concentrated solution of ammonium thiocyanate is added to a cobaltous solution, the latter becomes a beautiful blue, owing to the formation of ammonium cobaltous thiocyanate:

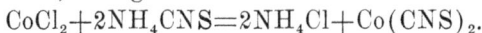

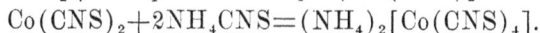

$$CoCl_2 + 2NH_4CNS = 2NH_4Cl + Co(CNS)_2.$$
$$Co(CNS)_2 + 2NH_4CNS = (NH_4)_2[Co(CNS)_4].$$

On adding water the blue colour disappears and the pink colour of the cobaltous salt takes its place. If, now, amyl alcohol is added (or a mixture of equal parts of amyl alcohol and ether), and the solution shaken, the upper alcoholic layer is coloured blue. This reaction is so sensitive that the blue colour is recognizable when the solution contains only 0.02 milligrams of cobalt. The blue solution also shows a characteristic absorption spectrum. Nickel salts produce no colouration of the amyl alcohol. If, however, iron is present, the red $Fe(CNS)_3$ is formed, which likewise colours the amyl alcohol, making the blue colour due to the cobalt, indistinct, so that, under some conditions, it cannot be detected. If, a little sodium carbonate solution or a few c.c. of concentrated ammonium acetate and 2 or 3 drops of 50 per cent. tartaric acid are added, the iron will be precipitated, the red colour produced by $Fe(CNS)_3$ will disappear, and the blue colour produced by the cobalt will be seen.

The above reaction serves as an excellent means of detecting cobalt in the presence of nickel.

Ether saturated with hydrochloric acid does not precipitate an anhydrous cobaltous salt, as in the case of nickel, but will dissolve the blue, anhydrous cobaltous chloride. This furnishes the basis of a method for separating nickel and cobalt.

a-Nitroso-β-naphthol, $C_{10}H_6(NO)OH$, produces a voluminous, purple red precipitate of cobalti-nitroso-naphthol, $[C_{10}H_6(NO)O]_3Co$, which is insoluble in cold, dilute nitric or hydrochloric acid.

This reagent serves not only for qualitative purposes, but can also be used for the quantitative determination of cobalt in the presence of nickel. The test may be applied conveniently to the solution obtained in the usual qualitative scheme after the removal of all metals except nickel and cobalt. A part of the solution may be used for the sensitive nickel test with dimethylglyoxime, and the remainder used for the cobalt test. To test for cobalt dilute the solution to about 50 c.c., add 4 c.c. of 6N hydrochloric acid and 20 c.c. of 6N acetic acid. Heat and add 50 c.c. of a saturated solution of nitroso-β-naphthol and boil in 50 per cent. acetic acid. If as much as 0.1 mg. of cobalt is present, a red precipitate or turbidity is obtained even in the presence of 250 mg. of nickel. When more than 150 mg. of nickel are present, however, some of the brownish-yellow nickel compound, $[C_{10}H_6(NO)O]_2Ni$, will precipitate after the solution cools.

The reagent used in this test should be freshly prepared. Nitroso-β-naphthol gradually decomposes on standing in the air, and changes from yellow to brown or even black in colour. It can be purified by dissolving in hot sodium carbonate, filtering, and reprecipitating with sulphuric acid. For ordinary purposes the saturated solution in 50 per cent. acetic acid is most suitable. The cobalt test can be made more delicate by adding an equal volume of alcohol to the test and, for detecting traces of cobalt, an aqueous solution of the organic substance can be used, but as 5,000 c.c. of water are required to dissolve 1 gram of the nitroso-β-naphthol, it is evident that the aqueous solution is not suitable when much cobalt is present. An excess of the reagent is required, as a part of it is used to oxidize the cobalt to the trivalent condition.

Copper gives a characteristic coffee-brown precipitate with the reagent, and it is possible to separate copper from lead, cadmium, etc., by means of it. Ferric iron gives a brownish-black precipitate which serves as a means of separating iron from aluminium, manganese, etc. Ferrous iron also gives a greenish precipitate in neutral solutions. Of all these precipitates, however, the cobalt compound is the most characteristic and the least influenced by the presence of acid. Thus with the acidity recommended above, the presence of a little ferric or ferrous iron causes no disturbance.

Reactions in the Dry Way

The bead produced by borax or sodium metaphosphate with cobalt salts is blue in both the oxidizing and reducing flames. By holding the bead in a reducing flame for a long time it is possible to reduce the cobalt to metal, when it appears, like nickel, gray.

On charcoal, cobalt compounds yield gray metallic cobalt, which can be removed by means of a magnet. The metal is placed on filter-paper, dissolved in hydrochloric acid and dried. The paper is then coloured blue by cobalt. If, now, sodium hydroxide is added and the paper exposed to the action of bromine vapours, black cobaltic hydroxide, $Co(OH)_3$, is formed.

Salts of cobalt when strongly heated with alumina give a blue-coloured compound, Thenard's blue, possibly $CoOAl_2O_3$.

Quantitative Determination of Cobalt and Nickel

In Oxidized Ores

The colorimetric method outlined below was used to determine cobalt and nickel in the ores of New Caledonia.

A 10 gram sample of the finely ground mineral was treated with hydrochloric acid and boiled to obtain complete decomposition and also to expel the chlorine which is set free in the reaction. By this treatment all the metals should be converted to soluble chlorides, leaving a residue of silica.

The solution was diluted to 100 c.c., the iron precipitated by powdered $CaCO_3$ or CaO, and the solution filtered. To the filtrate one or two drops of hydrochloric acid were added to remove any turbidity due to calcium carbonate. The clear solution was then poured into a standard colorimetric flask or tube and compared with standard colours. For the series of standard colours a solution of cobalt chloride was used. To prepare the cobalt chloride, a weighed quantity of cobalt nitrate was calcined, and the oxide obtained dissolved in concentrated hydrochloric acid. After the excess acid was removed, the cobalt chloride was dissolved and diluted to give solutions of the desired strength. Standard solutions containing from 0.25 to 5.0 per cent. cobalt oxide were prepared, the different solutions varying by 0.25 per cent. It was possible to determine the cobalt in New Caledonia ores colorimetrically to within 0.25 per cent.

Care must be taken in the preparation of colorimetric solutions to add a small quantity of nickel chloride (about one-third the contained cobalt chloride), for account must be taken of the nickel which the asbolite contains. The amount usually varies between one-half to one-third of the cobalt. Green nickel chlorides decrease a little the rose colouration of cobalt chloride, hence the addition of nickel chloride to the standards.

Electrolytic Determination of Cobalt and Nickel

In Oxidized Ores

The asbolite of New Caledonia contains silica, alumina, chromium, iron, manganese, nickel, and cobalt.

Thirty grams of finely pulverized mineral were treated with concentrated hydrochloric acid until decomposition was complete. In case the mineral was not completely decomposed, the hydrochloric acid solution was allowed to settle, and the clear liquid decanted through a filter. The residue was then heated with a little hydrochloric acid and a small quantity of nitric acid. After removing the excess acid, the solution was diluted and filtered. The residue was washed several times by decantation and finally it was brought into a filter and washed well with boiling water. The last portions filtered should be free from iron, the filtrate being tested by potassium ferrocyanide or sulphocyanide. The filtrates were combined and diluted to 900 c.c.

The residue contained the silica, silicates of alumina, chromite, etc.

To 150 c.c. corresponding to 5 grams of ore, a little sulphuric acid was added and the solution boiled. If any nitric acid was present the solution was evaporated to sulphuric fumes. If the evaporation was necessary, the solution was afterwards diluted to 100 c.c. In either case the iron was reduced by zinc or cadmium shavings, and titrated with standard potassium permanganate. Before titrating it was customary to test for any unreduced iron with potassium ferrocyanide.

From another 150 c.c., the iron, aluminium, and chromium were precipitated. The greater part of the acid was neutralized by sodium carbonate and then a quantity of ammonium chloride was added, followed by additions of barium or calcium carbonate. The solution was allowed to stand, with frequent stirring, to permit the oxide of chromium to be completely decomposed. After filtering, the precipitate was washed until the wash water did not show a precipitate on the addition of ammonium sulphide. The precipitate was dissolved in hydrochloric acid. From the diluted solution the iron, aluminium, and chromium hydroxides were precipitated by ammonia after the addition of several grams of ammonium chloride. This precipitation is repeated to remove any traces of barium or calcium carbonates.

The precipitate was dried and calcined. By subtracting the weight of iron found previously, the weight of the combined aluminium and chromium oxides was obtained. The precipitate was fused with sodium peroxide and the chromium determined by acidifying the sodium chromate solution, adding a standard ferrous sulphate solution and titrating with permanganate. The alumina was determined by difference.

From the 600 c.c. of solution, the iron, aluminium, and chromium were precipitated by calcium carbonate. After filtering and washing the precipitate, the solution was divided into two parts for the separation of manganese, nickel, and cobalt.

To one-half of the filtrate, which contained the manganese, nickel, and cobalt in 10 grams of ore, sodium carbonate was added in excess, then acetic acid to dissolve the precipitate. Thirty to fifty c.c. of sodium acetate were added to the solution which was afterwards saturated with hydrogen sulphide; the solution being kept at $70°C$. The precipitated sulphides of cobalt and nickel were removed by filtering and washed. The filtrate was tested for unprecipitated cobalt and nickel by additions of small quantities of ammonium sulphide which gives a black precipitate with cobalt and nickel salts. Ammonium sulphide was added to the filtrate to precipitate manganese sulphide, a flesh-coloured precipitate. The manganese sulphide was dissolved in hydrochloric acid and the manganese determined as manganese dioxide.

The precipitate of cobalt and nickel sulphides was dissolved in aqua regia and evaporated to dryness. A few c.c. of sulphuric acid were added and the solution heated to remove any volatile acids and destroy any organic matter. The solution of nickel and cobalt sulphates was diluted, neutralized with ammonia, and electrolyzed in a slightly acid solution.

To separate the cobalt and nickel, the metals were dissolved in nitric acid which was removed by evaporation and additions of hydrochloric acid. The almost neutral hydrochloric acid solution was saturated with chlorine gas or bromine, then an excess of calcium or barium carbonate was added. The solution was diluted and allowed to stand.

The cobalt hydrate was precipitated while the nickel remained in solution. Instead of chlorine or bromine, solutions of sodium hypochlorite or hypobromite may be used.

Another method (Rose's method) to separate the cobalt and nickel was to precipitate in a cold neutral solution, the cobalt hydrate by additions of barium or calcium carbonates in the presence of bromine water or chlorine gas. The use of barium carbonate and bromine was preferred. If the solution became acid, thus delaying and even stopping the precipitation, carbonate was added, the CO_2 expelled, and the solution cooled before adding the bromine water.

Zinc acts like carbonic acid, the smallest quantity retarding the precipitation. The composition of the precipitated black oxide was determined by dissolving in a mixture of HCl, adding potassium iodide and determining the iodine set free.

The above outline of the method of analyzing cobalt ores of New Caledonia was given by Beltzer.[1]

In Arsenical Ores

To 0.5 or 1.0 grams of ore add 10 c.c. of hydrochloric and 5 c.c. of nitric acid. After the action has ceased, add 20 c.c. of (1:1) sulphuric acid, evaporate to sulphuric fumes, and fume for five minutes. In most cases the treatment with hydrochloric and nitric acid may be omitted. To the cool sulphuric acid solution add 10 c.c. of cold water, then 15 c.c. of hydrochloric acid, sp. gr. 1.19 and heat on the hot-plate. Evaporate gently to fumes, then add 5 c.c. of nitric acid and the solution is again heated to fumes. After this operation there should be about 5 c.c. of H_2SO_4 present. About 90 per cent. of the arsenic is volatilized by this treatment.[2] The remainder of the arsenic and any copper are precipitated by H_2S from a hot solution containing 10 c.c. of hydrochloric acid in 100 c.c. of solution. The arsenic sulphide should be completely precipitated and coagulated in five to ten minutes. The solution is filtered using a 15 cm. No. 1 F. Swedish filter paper and the precipitate washed with hot H_2S water. The filtrate may be tested for unprecipitated arsenic by passing H_2S gas through the second time.

The clear filtrate is boiled to expel any H_2S, then 5 c.c. of hydrogen peroxide is added to oxidize the iron. Ammonia is added in excess to precipitate the iron. There is usually a sufficient quantity of ammonium salts present to prevent the precipitation of the cobalt and nickel. After the addition of ammonia, the solution is boiled, and the precipitate allowed to settle, filtered, and washed with hot water. As there is a tendency for the iron precipitate to retain some cobalt and nickel, the precipitate is dissolved in hot (1:4) H_2SO_4, diluted to 200 to 300 c.c., oxidized with hydrogen peroxide, reprecipitated with ammonia and treated as above. The combined filtrates are evaporated to 250 c.c. and electrolyzed. About 40 to 50 c.c. of ammonia should be present in the electrolyte. The addition of 0.5 grams of sodium sulphite before electrolyzing gives a better deposit. For each assay a current of 0.25 amperes running overnight is used with the ordinary stationary platinum electrodes. The best results are usually obtained when there is not more than 0.15 grams of metal on the cathode. With revolving anodes a much shorter time is required.

---

[1] Beltzer, La Chimie Industrielle Moderne, 1911.

[2] The writer is indebted to W. L. Rigg, chief chemist of the Deloro Smelting and Refining Company, for the outline of the above method of removing most of the arsenic by volatilization. The volatilization of arsenic chloride is not new, but the writer is advised that T. Melvor, a former chemist at Deloro, was the first to apply the method in the assays for cobalt and nickel.

When the cobalt-nickel solutions contain a large percentage of iron, it is advisable to precipitate cold and afterwards boil the solution. When using a revolving anode, it is not necessary to remove the iron before plating unless present in large quantity.

To test for any cobalt and nickel not deposited, remove a few c.c. of the electrolyte and add a few drops of ammonium sulphide. The formation of a black precipitate shows the presence of cobalt or nickel. The electrolyte may also be tested by removing a few c.c. and adding a few c.c. of dimethylglyoxime. Ammonium salts give a yellow colour, cobalt a brown colour, while nickel gives a red precipitate. The last of the nickel appears to be deposited after the cobalt.

After the deposition is complete the electrodes are washed and dried with alcohol and weighed.

The cobalt and nickel is removed from the electrode by placing it in a beaker containing about 7 c.c. of hot nitric acid. The electrodes are allowed to remain in the acid for 10 minutes, any undissolved metal being removed by tilting the beaker and turning the electrode in the acid. The electrode is washed with hot water, and the solution is evaporated to about 1 c.c. or less. Dilute to 10 c.c., neutralize with caustic potash and add sufficient hydrochloric acid from a pipette to dissolve any precipitate and four drops in excess. Then dilute to 50 to 150 c.c., depending on the quantity of nickel, (the larger volume with larger amounts of nickel), and boil the solution. To the hot solution add a sufficient quantity of a 1 per cent. alcoholic solution of dimethylglyoxime to combine with the nickel and cobalt. Nickel requires 5 and cobalt 1.7 times its weight of dimethylglyoxime. A few c.c. of the reagent in excess is necessary. When the solution is made slightly alkaline with ammonia a red precipitate of the nickel salt of dimethylglyoxime is formed. An excess of ammonia should be avoided, four drops are usually sufficient. After allowing the precipitate to stand for 30 minutes, filter into a weighed Gooch crucible. The precipitate is washed with hot water, dried at 105°C., and weighed. The weight of the precipitate multiplied by 0.2031 gives the weight of nickel. The weight of cobalt is found by difference. The dimethylglyoxime method of precipitating nickel is especially adapted for determining small amounts of nickel in the presence of large amounts of cobalt. It is satisfactory also when the nickel is high, but the quantity of solution taken should be such that there will not be more than 0.05 grams of nickel present.

The action of dimethylglyoxime on the nickel salt is as follows:

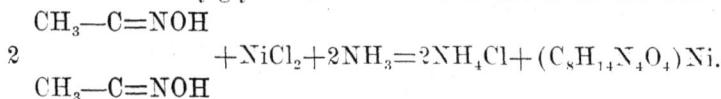

$$2 \begin{cases} CH_3-C=NOH \\ \\ CH_3-C=NOH \end{cases} +NiCl_2+2NH_3=2NH_4Cl+(C_8H_{14}N_4O_4)Ni.$$

According to L. Tschugaeff,[1] who first proposed this qualitative test, the presence of one part of nickel can be detected in the presence of 400,000 parts of water. The reaction is not influenced by the presence of ten times as much cobalt. When a larger proportion of cobalt is present the following procedure is adopted to detect traces of nickel in cobalt salts. Add strong ammonia to the solution of the cobalt salt until a clear solution is obtained, then add a few cubic centimetres of hydrogen peroxide and boil the solution a few minutes to decompose

---

[1] Berichte der Deutschen chemischen Gesellschaft, 1905, p. 2520.

the excess of this reagent. Then add the dimethylglyoxime and again bring the solution to a boil. A very small quantity of nickel causes a red scum to form and the sides of the beaker become coated with a film of red crystals. With smaller amounts of nickel the colour is best observed upon the filter through which the solution is poured and the precipitate being washed with hot water.

The above reaction is the most sensitive test known for detecting nickel in the presence of cobalt.

Separation of Cobalt from Nickel by Nitroso-β-Naphthol

Nickel and cobalt may also be separated by the use of Nitroso-β-Naphthol. This separation depends on the solubility of the nickel compound in hydrochloric acid, while the cobalt compound is insoluble.

To proceed with this method the metals are removed from the electrodes as above, 5 c.c. of sulphuric acid added, and the solution evaporated to sulphuric fumes. Cool, dilute and add 5 c.c. of concentrated hydrochloric acid. A freshly prepared hot solution of nitroso-β-naphthol in 50 per cent. acetic acid solution is added to the cobalt-nickel solution as long as a precipitate continues to form. The precipitate is allowed to settle and the solution tested for any cobalt. The solution is filtered, the precipitate washed, first with cold water, then with warm 12 per cent. hydrochloric acid to remove any nickel. Finally wash with hot water until free from acid. The precipitate is dried and heated strongly to convert the cobalt compound to oxide. After the carbon of the filter paper is all consumed, the cobalt is reduced to metal by heating in a current of hydrogen.

The disadvantages of the nitroso-β-naphthol method are, first, the precipitate cannot be converted to oxide and weighed because of the variable composition of the oxide, and second, the operation of reducing the oxide to metal requires more time and attention than the dimethylglyoxime method.

Separation of Cobalt and Nickel by Potassium Nitrite

After dissolving the cobalt and nickel in nitric acid, the solution is evaporated to a thick syrup. From 5 to 10 c.c. of water is added, and the solution neutralized by potassium hydrate. Any precipitate is dissolved in acetic acid, usually adding 8 c.c. of 1:1 acid in excess. The cobalt is precipitated as yellow tri-potassic-cobaltic nitrite by the addition of a 50 per cent. solution of potassium nitrite, the amount depending on the percentage of cobalt present. An excess of potassium nitrite is required, but usually 10 to 15 grams should be present in each assay. The solution is allowed to stand in a warm place for 15 hours, when the precipitate should have settled to the bottom of the beaker. The solution may be tested for unprecipitated cobalt by removing a portion, adding more nitrite, and allowing the solution to stand for an hour.

The solution is filtered and the precipitate washed with a 5 per cent. solution of potassium nitrite acidified with acetic acid, or with a 10 per cent. solution of potassium acetate, until the precipitate is free from nickel. The filter is removed from the funnel and spread on the inside of the beaker in which the precipitation was made. Most of the precipitate may be removed by a stream of hot water, while any remaining on the paper is dissolved by hot dilute sulphuric acid. The cobalt solution is evaporated to sulphuric fumes, cooled, diluted, made ammoniacal, and electrolyzed. The nickel is obtained by difference.

Separation of Zinc from Cobalt and Nickel

The usual method given to separate zinc from cobalt and nickel is to precipitate the two latter metals by hydrogen sulphide in a neutral, acetate, or formic acid solution.[1] Funk suggests neutralizing the solution with sodium carbonate and formic acid, and adding sodium formate equal to approximately three times the weight of the zinc. The quantity of cobalt and nickel present also has an effect on the purity of the precipitate and the completeness of the precipitation.

Zinc may be separated from cobalt by adding to a neutral solution potassium cyanide until any precipitate which forms is redissolved, then 10 c.c. of a 10 per cent. solution of potassium sulphide. After standing for 1-2 hours the zinc sulphide may be removed by filtration.

It is also possible to separate cobalt and nickel from zinc by electrolysis at 2.1-2.2 volts. With higher voltages the zinc is deposited.

Determination of Cobalt and Nickel in Cobalt Metal [2]

One gram of drillings contained in a 150 c.c. conical flask provided with a trap (made of a calcium chloride tube) is treated with 20 c.c. of hydrochloric acid, sp. gr. 1.12. A gentle heat is applied until all action ceases. One c.c. of nitric acid sp. gr. 1.4 is added to dissolve any remaining metallic residue. As soon as the action of the nitric acid is complete, the trap is rinsed, removed, and the solution evaporated to a syrup. The contents of the flask are taken up with 30 c.c. of water and filtered.

The siliceous residue on the filter is washed with water acidulated with a few drops of hydrochloric acid, incinerated, and fused with five times its weight of potassium pyrosulphate. The fusion is dissolved in a little water, and added to the main filtrate. The slightly acid solution of the metal is warmed and saturated with hydrogen sulphide. The sulphides of arsenic, copper, etc., are removed by filtration, thoroughly washed with acidulated hydrogen sulphide water (1 c.c. HCl: $100H_2O$) and the filtrate caught in a 350-c.c. casserole.

To expel the H_2S, the contents of the casserole are evaporated to a small volume. The iron is oxidized with a few drops of bromine, 0.2 grams of ammonium chloride added, and the evaporation continued to dryness at water-bath temperature.

The dry chlorides are dissolved in a little water, 0.1 gram of ammonium formate added, and the whole diluted to 50 c.c. The solution is heated until a precipitate of basic formate of iron separates. Very dilute ammonia is added until the solution is only slightly acid. After further heating for a few minutes, the precipitate of basic iron formate is allowed to settle, filtered, and washed with a hot dilute (0.1 per cent.) solution of ammonium formate.

The washed iron precipitate is dissolved off the filter with hot dilute (1:5) hydrochloric acid, the filtrate being caught in the casserole in which the iron precipitation was made. The solution of the iron precipitate is neutralized with ammonia, ammonium formate added, and the iron precipitation repeated in a volume of about 50 c.c. The precipitate is filtered and washed with the hot dilute ammonium formate solution as before.

The combined filtrates from the two iron separations are evaporated with

[1] These methods are summarized by Funk, Zeitschr. anal. Chemie, Vol. 46, 1907, pp. 93-106.

[2] Knittel, Can. Min. Jour., Vol. 36, 1915, p. 597.

the addition of 8 c.c. of concentrated sulphuric acid until fumes of sulphuric acid are copiously evolved.

The sulphates are dissolved in water and transferred to a 180 c.c. tall beaker, keeping the volume of the solution about 50 c.c. Sixty c.c. of ammonia, sp. gr. 0.9 is gradually added to the solution in the beaker (kept cool in running water) followed by 10 c.c. of 20 per cent. ammonium bisulphite solution.

The cobalt and nickel are deposited together with a current of 2.5 amperes. When the solution is colourless, the cover glass and the sides of the beaker are rinsed with water, and the current, reduced to 0.5 amperes, is allowed to pass until a few c.c. of electrolyte tested with potassium sulphocarbonate show that the cobalt and nickel are completely deposited. The cathode is removed with the usual precautions, dried, and the deposited cobalt and nickel weighed.

The cobalt and nickel are dissolved from the cathode with 30 c.c. of nitric acid (1:3), the cathode rinsed, removed, and the solution of the metals boiled to expel nitrous fumes. The solution is diluted to 500 c.c., neutralized with ammonia, made faintly acid with nitric acid, heated to about 50 to 60°C. and the nickel precipitated with a 1 per cent. alcoholic solution of dimethylglyoxime, followed by 10 c.c. of a 20 per cent. ammonium acetate solution.

The precipitate is allowed to stand for four hours, filtered on asbestos, washed twice with hot water, re-dissolved, and the precipitation repeated in a volume of 200 c.c.

After standing for an hour in a warm place, the nickel precipitate is filtered into a Gooch crucible, washed with hot water, and dried at 130 to 140°C. for forty-five minutes. The weight of the precipitate multiplied by 0.20316 gives the nickel. The amount of cobalt is found by difference.

Notes and Precautions.—Cobalt metal usually contains from 98 to 98.5 per cent. cobalt plus nickel. For this reason the amount of 0.2 to 0.3 grams of material recommended by some for the determination of the cobalt and nickel seems scarcely sufficient, as the weighing errors involved would appreciably affect the results. The use of large quantities of acids for solution and oxidation is to be condemned, as the removal of the excess consumes time and increases the chances of mechanical loss.

The separation of iron as basic formate is preferred on account of the ease with which it can be washed, and the formates are completely decomposed on evaporation with sulphuric acid.

The presence of acetates in the electrolyte seems to retard the complete deposition of the last traces of nickel. In one instance on electrolyzing a solution from metal containing 97.5 per cent. of cobalt and 0.8 per cent. of nickel, in the presence of acetates one milligram of nickel was found in the electrolyte 30 minutes after complete deposition of the cobalt. The volume of the electrolyte should be kept within the limit specified above, as the complete deposition of the metals from dilute solutions is unnecessarily prolonged.

It has been found that the amount of cobalt and nickel remaining in the electrolyte after electrolysis is less than 0.01 per cent. on a one gram sample.

The cathodes used are of the perforated type with an effective surface of 90 square centimetres. The anodes are spirals made of 0.04 inch wire, 0.6 inch diameter, and have about 6 turns.

Dry Assay for Nickel and Cobalt [1]

In this assay advantage is taken (1) of the facility with which nickel and cobalt may be concentrated in combination with arsenic to form a speiss; (2) of the order of oxidation of the metals which the speiss may contain, viz., iron, cobalt, nickel, and copper, and the colours they impart to borax. They are removed in the order named. Iron gives a brownish colour to borax, cobalt a blue, nickel a sherry-brown, and copper a blue. Hence, on scorifying the speiss with borax, the colour imparted by the oxide produced indicates the metal being removed. By careful and frequent examination of the colour resulting and the renewal of the borax it is possible to find the point at which first the iron and then the cobalt and nickel are removed. A greenish tint is imparted to the borax at the moment the cobalt begins to scorify, succeeded by a full blue (with fresh borax), followed by a greenish tint when the nickel commences to pass out. This changes to the full sherry-brown, and is followed by a greenish tint when copper commences to oxidize.

Careful examination and much care are necessary to obtain even fair results. By weighing the button at the various stages, the proportion of its constituents may be determined. If copper be present, 1 gram of gold is added to the button after the removal of the cobalt.

Assay of Ores and Speiss.

From 5 to 25 grams of the ore are finely powdered and passed through an 80-mesh sieve and calcined "sweet." At the end of the roasting some finely ground anthracite must be added, and the calcination continued till the carbon is burnt away, thus reducing the sulphates and arsenates formed in the earlier stages.

The roasted mass is mixed with 0.2 to 0.5 times its weight of arsenic, an equal weight of carbonate of soda, 5 grams of argol, and 2 to 4 grams of borax, melted in a crucible at a moderate temperature, and poured. If iron be absent, 0.5 grams of pure iron filings must be added before fusion.

When cold, the button is detached from the slag and weighed. It should be metallic in appearance, and have a smooth grey surface. Portions weighing 1 gram should be taken for the subsequent scorification.

The scorification with borax is conducted in small shallow dishes ¾ inch in diameter inside and ⅛ inch deep. These may be made of finely-sifted clay and ground pots. The clay should be stiff, and as much pressure as possible used in shaping them. The die may be made of boxwood, and provided with a gun-metal or iron ring. The dishes should be dried carefully and heated to dull redness in a muffle before use.

While preparing the speiss, a small muffle should be made as hot as possible, as the success of the operation depends largely on the temperature. The back of the muffle should be white-hot. Place a number of the small dishes in the muffle. Have at hand some ground borax glass, and a vessel of cold water. Place about a gram (rather less than more) of borax in one of the dishes as far from the front as can be seen. It is convenient to wrap the borax in tissue paper and drop in the speiss, also wrapped in tissue paper.

The muffle should be hot enough to melt the speiss immediately, or the order of oxidation will not be preserved. The borax should not be sufficient to cover the speiss when melted. For a moment the surface is dull, but almost instantly brightens and scorifies, very much like the brightening stage in the cupellation of silver. In a few moments remove the dish and contents, and immediately place the bottom of it in water to cool, and as soon as the bead is solid, submerge it in the water.

If iron only has passed off, the brownish-yellow tint due to that metal will only be observed, but if the smallest amount of cobalt has been removed the slag will be greenish or, if a larger quantity, blue. The correct stage has been reached when a faint green tinge is visible in the slags near the edge and round the button. If this be not observed, the operation is repeated till the point is reached. If it is past, the scorification is re-started with a fresh portion of speiss.

The speiss now only contains cobalt, nickel, and copper. It is weighed, and the operation repeated with every precaution till the cobalt is removed. Less borax is necessary as the bead is reduced in size, and a green cap of arsenate appears when the nickel commences to oxidize, as well as the greenish tinge in the slag near the bottom. The attainment of this point is marked also by the motion of the button momentarily ceasing. The process needs careful watching. The dish is withdrawn, and quenched carefully as before. If, on examination, it is doubtful whether the nickel has commenced to scorify, it is best to weigh the prill and return it to a scorifier with fresh borax, and examine immediately it is melted. The dense blue of the cobalt will not then interfere, and the brownish colour of the nickel (and the green cap) will be apparent. The prill is weighed. If copper were present in the speiss, the prill will now consist of nickel and copper arsenides. If much nickel is present the scorification may be continued in the same manner, but it is better to add 1 gram of pure gold, and continue the scorification so long as nickel continues to be removed. The resulting

[1] Assaying and Metallurgical Analysis, pp. 193-195, Rhead and Sexton; Longmans, Green and Company, 1911.

bead consists of the added gold and copper. It is weighed, and the increase in weight of the gold bead gives the copper. Confirmatory results may be obtained by cupelling the gold-copper bead with 34 times its weight of lead, when the gold only will be left, the loss of weight being copper.

In the above remarks it has been assumed that cobalt is present. If it is absent, it is difficult to ascertain the point at which iron is removed and nickel commences to pass out. Further, in assaying an unknown speiss, which may contain nickel and iron only, the green arsenate of nickel which forms on the surface and under the bead must not be confounded with the green tinge indicated above.

Modified Method.

In order to avoid the difficulty caused by the copper, it is sometimes removed before forming the speiss.

The sample of ore is digested with aqua regia till completely decomposed, hydrochloric acid added, and the nitric acid expelled by evaporation. Water is then added, and the liquor saturated with sulphuretted hydrogen, which precipitates the copper, etc. The liquid is filtered, and the residue washed with water containing sulphuretted hydrogen.

The filtrate is then boiled till sulphuretted hydrogen is completely expelled, oxidized by adding a few drops of nitric acid to the boiling solution and neutralized. To the neutral solution barium carbonate and bromine water are added in excess and well shaken. After boiling, the solution is filtered, and the precipitate washed, dried, and ignited. This precipitate, which contains the whole of the iron, cobalt, and nickel, is converted into a speiss as before, but without roasting.

Additional References

Aaron, Process of Precipitating Nickel and Cobalt from Solutions. The metals are precipitated as methyl sulphocarbonates; United States Patent, No. 330,454, Nov. 17th, 1885.

Grossmann and Schueck, Dycyandiamide in the Determination and Separation of Nickel. Engineering and Mining Journal, Vol. LXXXV, 1908, p. 1044.

Schoeller and Powell, The Determination of Nickel and Cobalt by the Phosphate method. The Analyst, Vol. XLI, 1916, pp. 124-131; Vol. XLII, 1917, pp. 189-199. Chem. Abst., Vol. XI, 1917, p. 2437.

Schoeller & Powell, The Determination of Cobalt and Nickel in Cobalt Steel, Jour. Iron & Steel Inst., Vol. XCVII, No. 1, 1918, pp. 441-449.

Powell, The Estimation of small quantities of Cobalt. Jou. Soc. Chem. Ind., Vol. XXXVI, 1917, pp. 273-274.

Walker, Separation of Nickel and Cobalt by Red Lead. Eng. Min. Jour., Vol. 103, 1917, p. 894.

The Use of Dimethylglyoxime as an Indicator in the Volumetric Determination of Nickel by Frevert's Method, Jour. Ind. Eng. Chem., Vol. 8, 1916, pp. 804-807.

Metzl, The Volumetric Estimation of Cobalt in the Presence of Nickel, Zeitschr. Anal. Chemie, Vol. 53, 1915, p. 537.

CHAPTER IV

THE USES OF COBALT

Cobalt Oxide

Cobalt is used chiefly in the form of oxide in the enamel, porcelain, and glass industries, but within the last few years new uses have been found for the metal which is at present produced in considerable quantity. Cobalt metal is used chiefly in the manufacture of stellite, a cobalt-chromium alloy, used as a cutting tool. The metal is added to some high-speed steels to give improved cutting qualities. It is also used in cobalt plating.

Cobalt oxide and its compounds are used as pigments or colouring agents. It is said that when cobalt oxide is present in the ratio of 1:20,000, it imparts a bluish tinge to clear glass or porcelain. The oxide is black or gray, but when fused with borax or silica it possesses a brilliant blue colour. Cobalt oxide is also used in small proportions to produce white enamels, since any yellow colour due to iron oxide is neutralized by the complementary cobalt blue, producing a pure white. Also by the addition of cobalt oxide, copper oxide, pyrolusite, and even iron oxide, to certain raw mixtures or waste enamels, a beautiful black enamel is obtained. The compounds of cobalt, for example, silicate, aluminate, phosphate, arsenate, and nitrite are used instead of the oxide, because they give better and more uniform colouring. The following table gives a list of the customary brands of cobalt compounds with their cobalt content:—

Brand.	Special Designation.	Chemical Formula.	Percentage Cobalt Content.
F F Ko	Finest cobalt oxide (superior oxide)	CoO	78* per cent.
G K O	Grey cobalt oxide, Ia	CoO	76* "
F K O	Grey cobalt oxide	CoO	75* "
R K O	Black cobalt oxide, Ia	Co_2O_3	70* "
S K O	Black cobalt oxide	Co_3O_4	66* "
A K O	Cobalt arsenate	$Co_3As_2O_8.8H_2O$	29 "
K O H	Cobalt carbonate	$CoCO_3$	50 "
P K O	Cobalt phosphate	$Co_3(PO_4)_2.8H_2O$	34 "

* Theoretically CoO, Co_2O_3, and Co_3O_4 contain 78.8, 71.1, and 73.4 per cent cobalt, respectively.

The history of the value of cobalt compounds as colouring agents dates back to pre-historic times. However, it may be stated that it was not until the discovery of the silver-cobalt deposits at Schneeberg in 1470, that cobalt was used to any great extent. The preparation of cobalt compounds must have been carried on in a small way because about the year 1790, there were 25 works engaged in the industry, most of which were located in Saxony, and the total production of these works was not more than 300 tons of cobalt annually, which was mostly in the form of smalt. The smalt which contained approximately 6 per cent. cobalt was sold in Venice in 1520 at about 16 cents a pound. There were also a few refineries in Holland which supplied the Irish linen trade almost entirely, as well as a large amount to the linen industries at home. It was also used in Holland in the manufacture of litmus. A complete description of the early history of the cobalt industry in Saxony, is given by Mickle in the Report of the Bureau of Mines of Ontario, vol. XIX, 1913, Pt. II, pp. 234-251.

At present the ceramic industry is carried on chiefly in the United States, Germany, France, and Austria-Hungary. In Germany and Austria-Hungary it gives employment to 50,000 people.

Smalt is used now only in a few enamel works. It is a blue compound which owes its colour to the presence of cobalt silicate. As formerly prepared it contained appreciable quantities of impurities. The oxides of cobalt are preferred to smalt because of their purity, uniformity, and lower cost.

The arsenate is prepared by adding sodium arsenate to a cobalt nitrate solution.

Cobaltous carbonate is obtained by adding soda or potash to a solution of a cobalt salt. The rose coloured precipitate which forms is a basic carbonate, of the formula $CoCO_3 + Co(OH)_2$.

Cobalt phosphate is prepared by adding sodium phosphate to a cobalt acetate solution. The precipitate is violet in colour and has the formula $Co_3(PO_4)_2$.

The aluminate is formed by adding sodium carbonate to a mixture of cobalt nitrate and alum. The cobalt and aluminium hydroxides may be precipitated separately and afterwards mixed. The mixed hydroxides are washed, dried, and heated at a red heat. The blue cobalt aluminate which forms is ground and dried.

The colour produced by the aluminate, phosphate, or arsenate has various names, for example, cobalt blue, cobalt ultramarine, king's blue, Thenard's blue, or azure blue. Thenard's blue corresponds to cobalt aluminate. Coeruleum, coeline, or blue celeste, is a blue colour showing a slightly greenish tint. It contains oxide of tin and sometimes calcium sulphate. To prepare such a pigment, sodium stannate is added to a cobalt nitrate solution. The precipitate is washed and heated. Another method to prepare blue celeste is to heat cobalt sulphate, tin oxide, and precipitated silica or chalk.

Mazarine blue [1] is commonly employed as a band on the edges of plates. The colour is prepared by mixing cobalt oxide, with tin oxide, sand, and calcium sulphate.

New blue is a pigment varying in colour from a pale greenish blue to a deep turquoise blue. It is largely used for enamels, and consists of aluminates of cobalt and chromium produced by the action of alum on carbonates and hydrates of cobalt and chromium.

Cobalt green or Rinmann's green is formed by substituting zinc oxide for alumina in cobalt aluminate, giving cobalt zincate. This compound may also be formed by mixing the hydroxides or oxides or by adding soda to a cobalt-zinc solution. In either case the oxides must be heated to form the zinc compound. The darker green colours contain the smaller quantities of zinc. A mixture of calcined cobalt carbonate, chromium oxide, and alumina also produces a green pigment.

Cobalt bronze is a cobalt ammonium phosphate compound. It has a violet colour with a bronze-like metallic lustre.

Cobalt yellow, Indian yellow, aureolin, is the precipitate potassium cobaltic nitrite. It is prepared by adding potassium nitrite to a cobalt solution acidified with acetic acid. It is a bright yellow precipitate which because of its purity produces an excellent colour.

---

[1] Mellor, Clay and Pottery Industries, 1914, p. 71, Lippincott, Philadelphia.

Cobalt brown is formed by calcining a mixture of ammonium sulphate, cobalt sulphate and ferrous sulphate.

In the burning operation of cobalt compounds, it is important that the temperature should not be too high, as a high temperature produces unsatisfactory colours. Of the cobalt compounds the silicate, carbonate and phosphate are the most important.

Cobalt oxide, up to 0.5 per cent., is used in practically all ground enamels as it possesses the property of causing the enamel to adhere better to sheet iron and at the same time neutralizes any yellow colour due to iron oxide. Although numerous investigations have been undertaken to account for this property, no satisfactory explanation has yet been given.

Blue enamels contain on an average 1 per cent. of cobalt, but when a dark blue colour is desired, cobalt may be present up to 3 per cent.

Red and pink cobalt compounds are of scientific rather than technical interest. If cobalt arsenate is strongly heated and then ground it yields a pinkish-red powder. The precipitate obtained from a solution of a cobalt salt with sodium phosphate is pink, changing to violet when heated. Cobalt magnesia pink is obtained from precipitated magnesium carbonate, mixed to a thin paste with cobalt nitrate solution, dried, and heated in crucibles.

Sympathetic inks.—Many of the salts of cobalt are pink and deliquescent. If a weak aqueous solution of one of them, such as the nitrate or chloride, is used as ink, the writing is practically invisible, but if the paper is held near the fire the combined water is driven off and the writing becomes blue and visible. It will afterwards absorb water from the atmosphere and again disappear.

A few experiments have been made to test the action of cobalt nitrate as an addition agent in flotation but it did not show any advantages.[1]

Uses of Metallic Cobalt

Metallic cobalt is used chiefly in the preparation of alloys and high-speed alloy steels. The cobalt-chromium alloys are the most important. These alloys possess extreme hardness and are being used extensively to replace high-speed steels as cutting tools. The trade name of the cobalt chromium alloys is " stellite." Stellite is not a steel, and its properties are altogether different from those of steel. It cannot be hardened or tempered, nor does it lose any of its hardness even when the edge of the tool is at a red heat. Tests have shown that a stellite cutting tool permits more rapid cutting than when the ordinary high-speed steel is used.

The cobalt-chromium alloys are hard, but the hardness is increased by additions of tungsten and molybdenum. As the hardness increases the brittleness also increases. The addition of iron softens the alloy.

Stellite alloys possess a bright surface, and are very resistant to oxidation. They remain unaltered in the atmosphere, and are not attacked by the ordinary acids. The colour of the alloys, when polished, lies between that of steel and silver.

[1] Metallurgical and Chemical Engineering, Vol. XVIII, 1918, p. 76.

The tools are made by casting the alloy into bars of the desired shape and size. These are afterwards ground to a cutting edge on an emery or carborundum wheel.

Two grades of stellite tools are made, one with moderate hardness and great strength for turning steel, and the other with greater hardness but less strength for turning cast iron. The tool used for cast iron enables a greater amount of work to be accomplished, whereas, if operating at a high speed with the stellite tool for steels, the edge would be immediately destroyed.

Further information on these alloys will be found in the section on alloys.

The use of stellite alloys for cutlery has been suggested but up to the present, it has not been used for this purpose as the demand for the cutting tool has been so great.

Electro-plating with Cobalt

Owing to the success that has attended nickel plating, the question arose as to whether cobalt platings possess any superior qualities to nickel platings. In order to decide this question a number of experiments were undertaken at the School of Mining, Kingston, Canada, for the Mines Branch of the Department of Mines. A report[1] of this investigation has been issued, and in it some interesting conclusions are given. The results of the work were tested and confirmed by disinterested operators.

The advantages claimed for cobalt plating may be summarized as follows:—

1. Cobalt may be plated from four to fifteen times as quickly as nickel.

2. The cobalt plating is harder than the ordinary nickel plating.

3. About one-fourth the weight of cobalt as compared with nickel is required to do the same protective work.

Cobalt may be plated on brass, iron, steel, copper, tin, German silver, lead and Britannia metal.

The composition of the solutions recommended is as follows:—

Solution 1 B.—Cobalt-ammonium-sulphate, $CoSO_4.(NH_4)_2SO_4.6H_2O$—200 grams to the litre of water, which is the equivalent of 145 grams of anhydrous cobalt-ammonium-sulphate, $CoSO_4.(NH_4)_2SO_4$, to the litre of water. Sp. gr.= 1.053 at 15°C.

Solution XIII B:—Cobalt sulphate, $CoSO_4$—312 grams; sodium chloride, NaCl—19.6 grams; boric acid—nearly to saturation; water—1,000 c.c. Sp. gr.= 1.25 at 15°C.

Further experiments are being conducted to test the value of cobalt platings.

Kowalke[2] gives an account of a few experiments made to test the suitability of cobalt for use in thermocouples. He states that cobalt should have an important place among thermo-elements since it does not become brittle like nickel, and it gives a high electromotive force.

An amalgam of cobalt is used in dentistry.

---

[1] Kalmus, Electro-plating with Cobalt: Bulletin No. 334, Department of Mines, Ottawa, 1915. Trans. Am. Electrochemical Society, Vol. XXVII, 1915, pp. 75-130.

[2] Cobalt as an Element for Thermocouples, Trans. Amer. Electrochem, Soc., Vol. XXIX, 1916, pp. 561-568.

A French patent (No. 460,093, July 7, 1913) covers the preparation of cobalt filaments for incandescent electric lamps. The filament is made from a solution of cellulose with zinc chloride, cobalt oxide, and manganese sulphate. It is heated to incandescence for twenty hours and then coated with carbon.

Additional References

Mellor, Cobalt Blue Colours, Trans. Eng. Ceramic Soc., Vol. VI, 1907, p. 71.

Use of Cobalt in Decorating, Chem. Abst., Vol. VII, 1913, p. 3003.

Why a Greater Colour is produced with Cobalt Solutions than with Mineral Colours, Chem. Abst., Vol. VII, 1913, p. 554.

Old and New Colours with Cobalt as their Base, Chem. Abst., Vol. VII, 1913, p. 874.

The Necessity of Cobalt Oxide in Ground-coat Enamels for Sheet Steel, Trans. Am. Ceramic Soc., Vol. XIV, 1912, pp. 756-764.

Cobalt Uranium Colours, Chem. Abst., Vol. VIII, 1914, p. 3710.

Status of Cobalt in the Ground-coat of Sheet Steel Enamels, Chem. Abst., Vol. VIII, 1914, p. 3847.

Cobalt in Pottery Decoration, Chem. Abst., Vol. VIII, 1914, p. 798.

Cobalt Oxides, Reactions between CoO and Al_2O_3, Chem. Abst., Vol. IX, 1915, p. 2852.

Cobalt Oxides, Reactions between CoO and SnO_2, Chem. Abst., Vol. IX, 1915, p. 2844.

Cobalt Magnesium red, Chem. Abst., Vol. IX, 1915, p. 2853.

Cobalt Colours other than Blue, Trans. Am. Ceramic Soc., Vol. XIV, 1912, pp. 767-777.

The Formation of Isomorphous Mixed Crystals between Cobalt Oxide and Manganese Oxide and between Cobalt Oxide and Nickel Oxide, Chem. Abst., Vol. IX, 1915, p. 3039.

Hedvall, The Determination of Dissociation Temperatures with the Aid of Cooling and Heating Curves, Especially for Cobalto-cobaltic Oxide, Chem. Abst., Vol. XI, 1917, p. 1347.

Hedvall, The formation of Cobalt Aluminate, Cobalt Orthostannate and Rinmann's Green, Chem. Abst., Vol. XI, 1917, p. 1373.

CHAPTER V

BINARY ALLOYS OF COBALT

Cobalt and Aluminium

The equilibrium diagram of the cobalt aluminium alloys[1] is shown in Figure. 1. The liquidus curve[2] consists of 4 branches, viz.: AB, BC, CDE, and EF. The points A and F, shown at 658° and 1492°C., correspond to the melting points of aluminium and cobalt respectively. The point E at 1375°C. and 90.5 per cent. cobalt is a minimum point. Alloys of composition represented by points to the right of E consist when solid of a solid solution of aluminium in cobalt. The point D at 1628°C. and 68.5 per cent. cobalt corresponds to the melting point and composition of the compound CoAl. At C, 1165°C. and 38 per cent. cobalt, there is a reaction between the CoAl crystals and the liquid of composition C to form a new compound Co_2Al_5, containing 46.5 per cent. cobalt. At B, 940°C. and 20 per cent. cobalt, there is a reaction between the previously formed Co_2Al_5 crystals and the liquid of composition B to form the compound Co_3Al_{13}, containing approximately 33.5 per cent. cobalt.

The alloys containing between 100 and 68.5 per cent. cobalt are magnetic, the magnetism decreasing rapidly with increased aluminium content.

Schumeister[3] has studied the mechanical and chemical properties of aluminium cobalt alloys containing from 0 to 20 per cent. cobalt. The tests show that the alloys containing 9 to 12 per cent. cobalt possess the greatest tensile strength. Addition of small quantities of tungsten, 0.8 to 1.2 per cent. raised the tensile strength considerably, while with further additions the strength is lowered. The substitution of molybdenum for tungsten did not show any advantage. The aluminium cobalt alloys were harder, easier to work, more stable and durable in the air than pure aluminium.

Peltery obtained a patent[4] on the addition of silver, gold, cobalt, chromium, iron, manganese, and nickel to aluminium; also the General Electric Company, Berlin, patented a light bearing metal containing aluminium with lead, tin, cobalt, chromium, iron, molybdenum, nickel, and antimony.[5]

The following table gives the results obtained later by Schumeister[6] from additions of cobalt, 0 to 12 per cent., on the tensile strength, elongation, and hardness of aluminium. The fracture of the alloys changes from a coarse to a very fine grain with increased proportion of cobalt. The following table shows the effect of additions of cobalt to aluminium.

---

[1] Gwyer, Aluminium and Cobalt: Zeitschr. anorg. Chemie, Vol. LVII, 1908, pp. 140-147.

[2] The liquidus or freezing point curve represents the beginning of freezing or solidification of any alloy.

[3] Schumeister, Investigation of the Mechanical and Chemical Properties of Light Cobalt Aluminium Alloys: Metallurgie, Vol. VIII, 1911, pp. 650-655.

[4] German Patent, No. 230,095, Jan. 16th, 1911.

[5] German Patent, No. 257,868, March 20th, 1913.

[6] Schumeister, Investigation of the Binary Aluminium Alloys: Stahl and Eisen, Vol. XXXV, 1915, pp. 873, 874.

7 B.M. (iii)

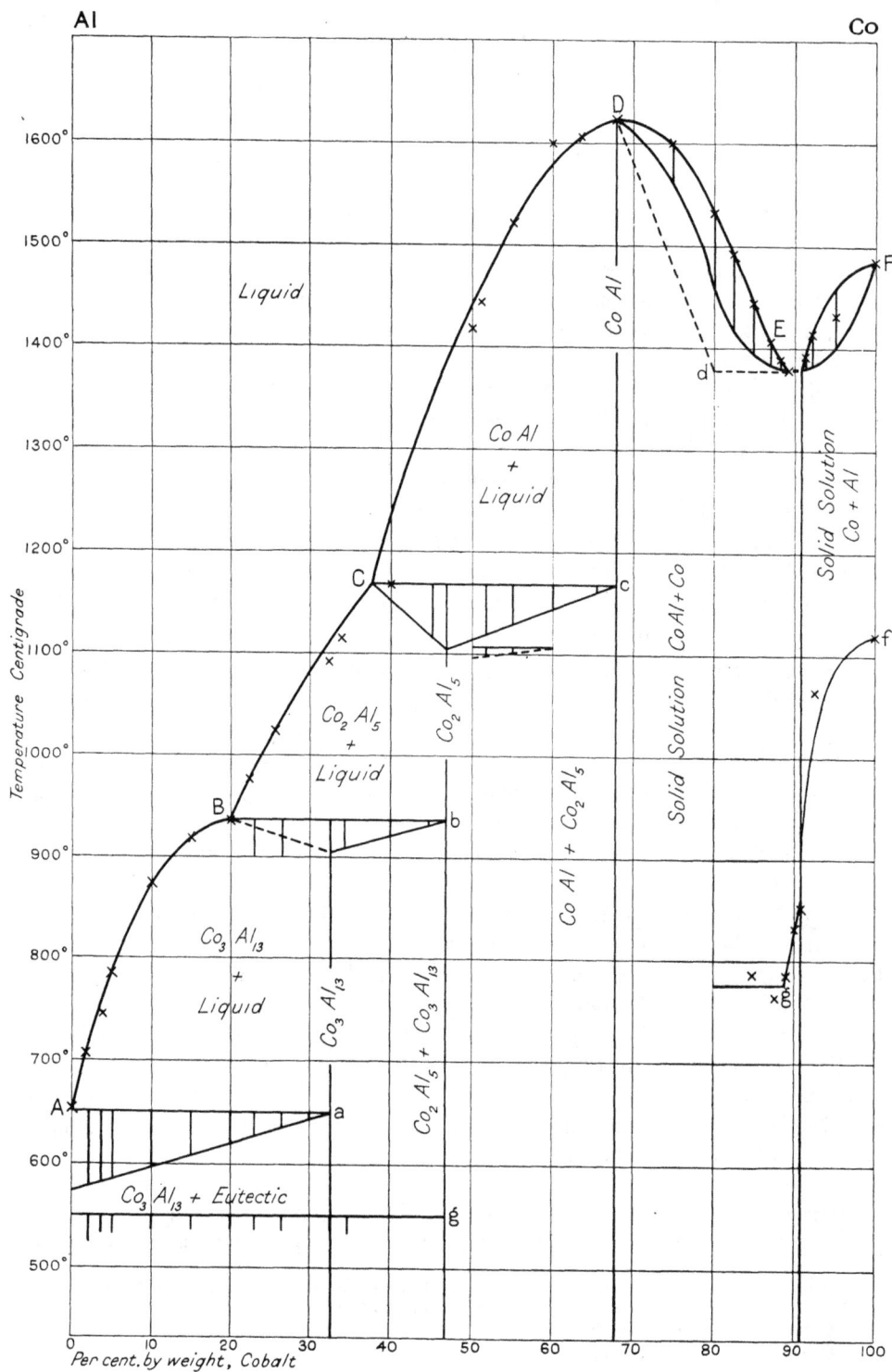

Figure 1.—Equilibrium Diagram of Aluminium-Cobalt Alloys.

Cobalt Content	Tensile strength per sq. m.m.	Elongation per cent.	Hardness
0.0	10.5	34	29
0.6	10.9	35	32
1.6	12.0	28	....
2.3	12.3	25	....
3.5	12.9	21	47
5.5	15.5	18	....
7.5	16.6	14	50
9.4	16.5	11	51
10.5	17.0	11	....
12.0	18.5	6	61

Additional References

Portevin, Aluminium Alloys: Revue de Metallurgie (Mémoires), Vol. V, 1908, p. 274.

Bornemann, Cobalt and Aluminium: Metallurgie, Vol. VII, 1910, pp. 577, 578.

Kaiser, Metallurgie, Vol. VIII, 1911, p. 300. Analysis of White Bronze (Cu 40-30, Co 50-60, Al 10); p. 305, Analysis of Metalline (Cu 30, Co 35, Al 25, Fe 10).

Cobalt and Antimony

The equilibrium diagram[1] of the cobalt antimony alloys is shown in Figure 2. Both metals are soluble in one another in the liquid state, but only to a small extent in the solid. At 1093°C., cobalt retains 12.5 per cent. by weight of antimony in solid solution; the amount, however, decreases with the temperature. There is no evidence of antimony dissolving cobalt in the solid state.

The addition of antimony lowers the melting point of cobalt, A, until the eutectic point E is reached at 1093°C. and 39 per cent. antimony. The liquidus curve shows a maximum C at 1191°C. and 67 per cent. antimony, corresponding to the compound CoSb. At 897.5°C. there is a reaction between the separated crystals CoSb and the liquid D to form a new compound F. The exact composition of the compound F has not yet been definitely determined. From the eutectic point G, at 615°C. and 98.5 per cent. antimony, the liquidus rises to H at 630°C. the melting point of antimony.

The transformation temperature of cobalt at 1159°C. is lowered to 930°C. by addition of antimony up to 12.5 per cent. For all alloys between 12.5 and 67 per cent. Sb, the transformation takes place at constant temperature, viz., 930°C.

Additional References

Lewkonja, Cobalt-Antimony Alloys: Zeitschr. anorg. Chemie, Vol. LIX, 1908, pp. 305-312.

Ducelliez, Alloys of Cobalt and Antimony: Procès verbaux de la société des sciences physiques et naturelles de Bordeaux, 1908, pp. 183-190; 1908-1909, pp. 131-134. Bulletin Société Chimique de France, Vol. VII, 1910, sec. 4, p. 202.

Ducelliez, Action of Antimony Chloride on Cobalt and its Alloys with Antimony: Compt. Rend., Vol. 147, 1908, pp. 1048-1050.

Ducelliez, Study of the Electromotive Force of the Alloys of Cobalt with Tin, Antimony, Bismuth, Lead, and Copper: Compt. Rend., 1910, Vol. 150, pp. 98-101.

Kurnakow and Podkapajew, Jour. russ. phys. Chem. Ges. 38, 1906, p. 463.

Rammelsberg, Annalen der Physik und Chemie, Vol. 128, 1866, p. 441.

Bornemann, Cobalt and Antimony: Metallurgie, Vol. VIII, 1911, pp. 683-684.

---

[1] Guertler, Metallographie, Vol. I, 1912, pp. 754-756.

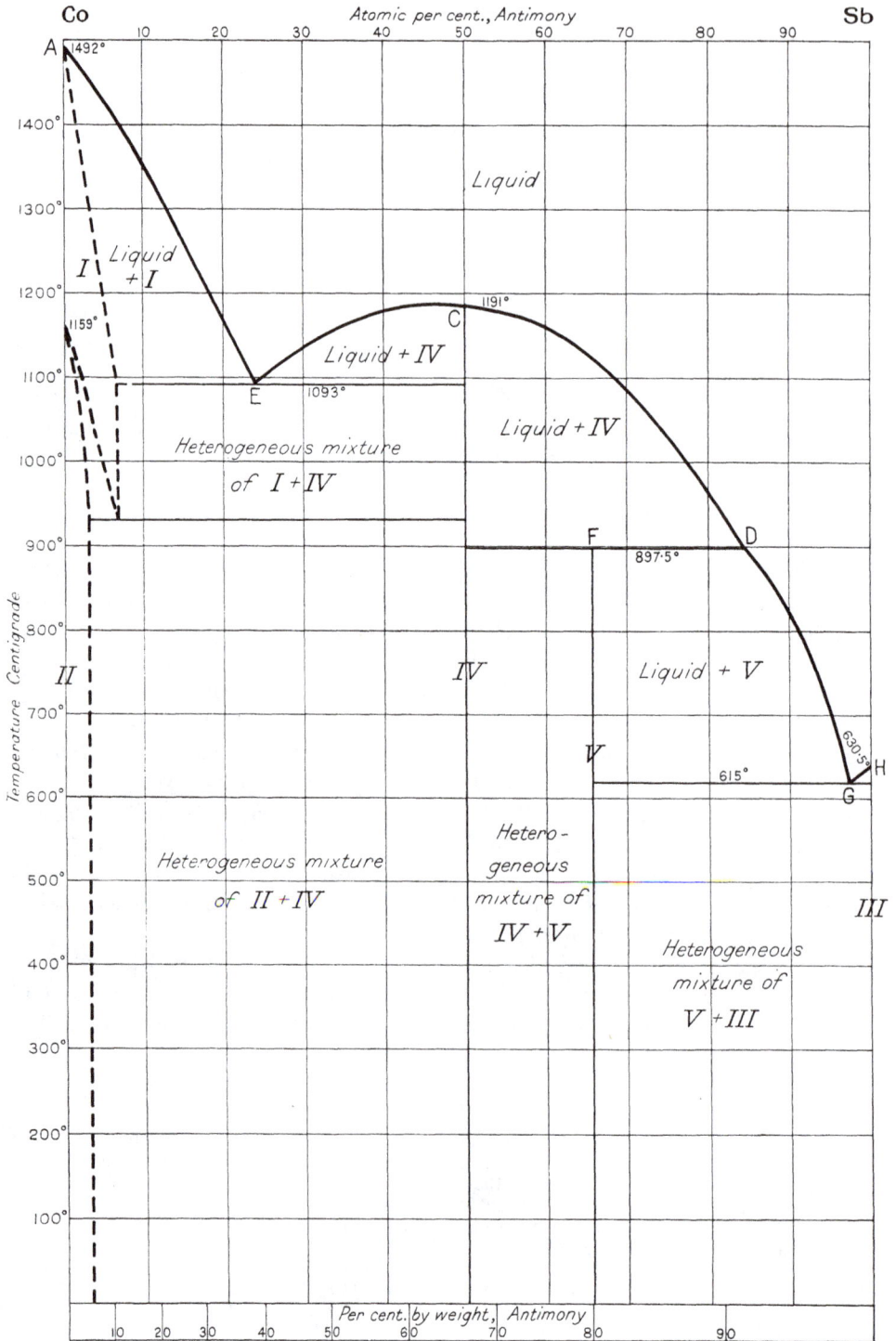

Figure 2.—Equilibrium Diagram of Cobalt-Antimony Alloys.

Cobalt and Arsenic

The cobalt arsenic equilibrium diagram[1] is shown in part in Figure 3. Owing to the volatilization loss of arsenic it has been impossible to investigate concentrations of more than 60 per cent. arsenic.

Cobalt retains about 1 to 2 per cent. arsenic in solid solution. At a temperature of 920°C. and a concentration of 30 per cent. arsenic, there is a eutectic point. With further additions of arsenic, the liquidus curve rises in several successive stages to a maximum at 1175°C. and 57 per cent. arsenic. This maximum corresponds to the compound CoAs. At the temperatures 1014°, 960°, and 930°C., there are three changes in the direction of the liquidus corresponding to the separation of crystals IV, V, VI, of the composition approaching Co_3As_2, Co_2As, and Co_5As_2. Conclusive evidence of the existence of these compounds has as yet not been obtained.

There are also three horizontals in the diagram, at 910°, 830°, and 380°C. The horizontal at 910° between crystals V and III, shows a change in the solid state of crystals IV to crystals VII of similar composition. The horizontal at 830° reaching between crystals I and VI corresponds to a change in the solid crystals VI to IX of the same composition. For compositions containing mixtures of crystals V and VII there appears to be a transformation at 380°, but positive evidence of the exact nature of this change is lacking.

Additional References

Friedrich, Equilibrium Diagram of the Cobalt-Arsenic Alloys: Metallurgie, Vol. V, 1908, pp. 150-157.

The Freezing-point of Cobalt-Nickel Arsenides: Metall und Erz, Vol. X, 1913, pp. 659-671.

Ducelliez, Action of Arsenic Chloride and Arsenic on Cobalt: Compt. Rend., Vol. CXLVII, 1908, p. 424.

Action of Heat on Mixtures of Arsenic and Cobalt: Procès verbaux de la société des sciences physiques et naturelles de Bordeaux, 1908, pp. 57-73.

Cobalt and Bismuth

The equilibrium diagram[2] of the cobalt bismuth system is shown in Figure 4. The metals are only partly soluble in the liquid state. At 1390°C. the concentration of the two layers is 92.7 per cent. cobalt and 7.3 per cent. bismuth; and 93 per cent. bismuth and 7 per cent. cobalt. The first addition of bismuth to cobalt lowers the melting point of the latter by approximately 100°C. The addition of cobalt to bismuth lowers the melting point of bismuth approximately 10° to a eutectic point at 96.7 per cent.

Additional References

Lewkonja, Cobalt-Bismuth Alloys: Zeitschr. anorg. Chemie, Vol. LIX, 1908, pp. 315-318.

Ducelliez, Study of Electromotive Force of Cobalt-Bismuth Alloys: Bulletin Société Chimique de France, Vol. VII, 1910, pp. 199-200.

Cobalt-Bismuth Alloys: Bulletin Société Chimique de France. Vol. V, 1909, pp. 61-62.

Bornemann, Cobalt and Bismuth: Metallurgie, Vol. VIII, 1911, p. 688.

[1] Guertler, Mètallographie, Vol. I, 1912, pp. 833-836.
[2] Guertler, Metalloghaphie, Vol. I, 1912, pp. 584-586.

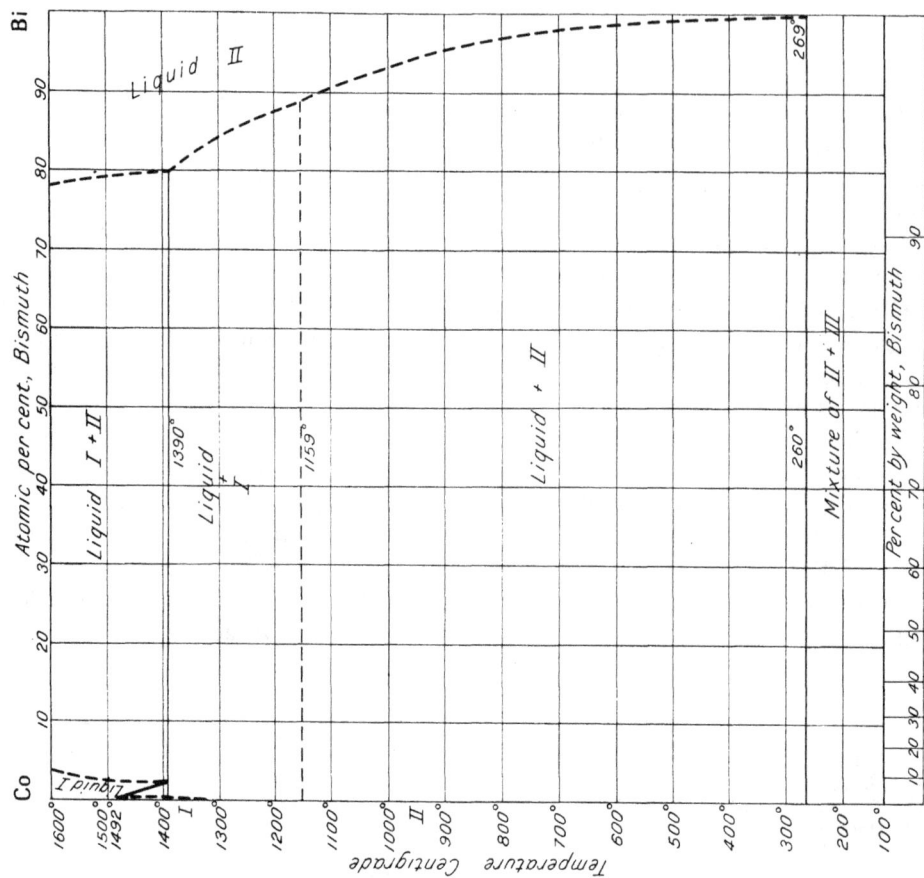

Figure 4.—Equilibrium Diagram of Cobalt-Bismuth Alloys.

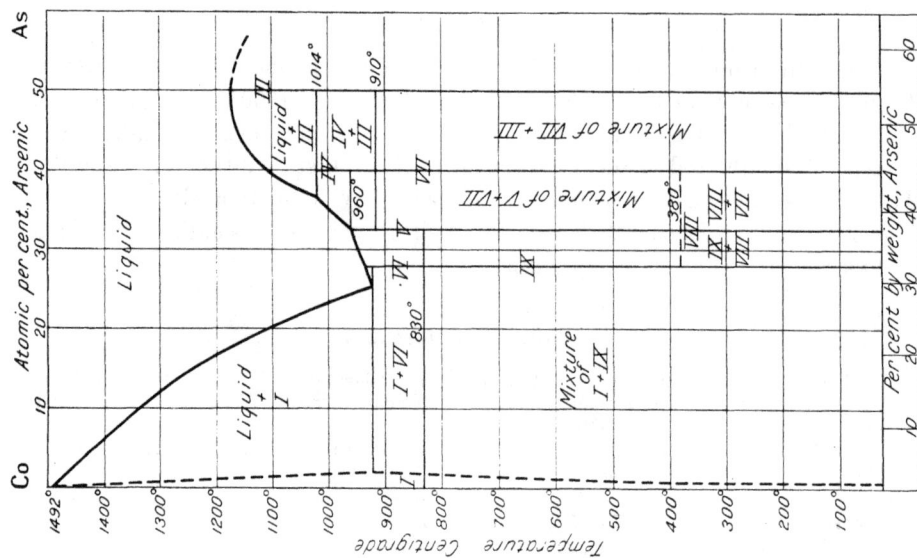

Figure 3.—Equilibrium Diagram of Cobalt-Arsenic Alloys.

Cobalt and Boron

The equilibrium diagram of cobalt and boron alloys or mixtures has not yet been published. However, the compounds Co_2B and CoB_2 have been detected. . The magnetic transformation of Co_2B occurs at 156°C.

References

Jassonneix, The Combination of Nickel and Cobalt with Boron: Compt. Rend., Vol. CXLV, 1907, pp. 240-241.

A Study of the Magnetic Properties of Iron, Cobalt, Nickel, and Manganese with Boron. Report of Eighth International Congress of Applied Chemistry, Vol. 2, 1912, pp. 165-170.

Moissan, A Study of the Borides of Cobalt, Compt. Rend., Vol. CXXII, 1896, pp. 424-426.

Cobalt and Cadmium

A few experiments on the addition of cobalt to cadmium were made by Lewkonja.[1] In all the tests a distinct eutectic point was observed at 316°C., 6° below the melting point of cadmium. None of the alloys were magnetic. It is evident that there must be either a compound formed, or that a solid solution of cadmium in cobalt must exist, as the transition to a cobalt was lowered to below the room temperature.

Additional Reference

Guertler, Metallographie, Vol. I, 1912, p. 487.

Cobalt and Carbon

The equilibrium diagram of the cobalt carbon series has been investigated by Boecker,[2] and later by Ruff and Keilig.[3] The diagram of Ruff and Keilig is shown in Figure 5.

Both diagrams show the existence of a eutectic at 1300°C. and 2.8 to 2.4 per cent. carbon. At the eutectic temperature cobalt retains 0.82 per cent. carbon in solid solution, which separates as graphite on cooling, only 0.3 per cent. being retained at 1000°C. Boecker did not carry the experiments above 1700°C., at which temperature he found the maximum solubility of carbon in cobalt to be 3.9 per cent.

The investigations of Ruff and Keilig were conducted at temperatures up to 2415°C., which is the boiling point of the liquid under 30 m.m. pressure. The boiling point of pure cobalt was found to be 2375°C. under these conditions. The existence of a cobalt carbide has not yet been definitely proved.

Additional References

Keilig, The Cobalt-Carbon System for Temperatures above 1500°C. Dissertation, Königliche Technische Hochschule in Danzig, 1915.

Guertler, Metallographie, Vol. 2, 1913, p. 639.

---

[1] Lewkonja, Cobalt-Cadmium Alloys: Zeitschr. anorg. Chemie, Vol. LIX, 1908, p. 322.

[2] Boecker, Investigation of the Cobalt-Carbon System: Metallurgie, Vol. IX, 1912, pp. 296-303.

[3] Ruff and Keilig, Cobalt and Carbon: Zeitschr. anorg. Chemie, Vol. LXXXVIII, 1914, pp. 410-423.

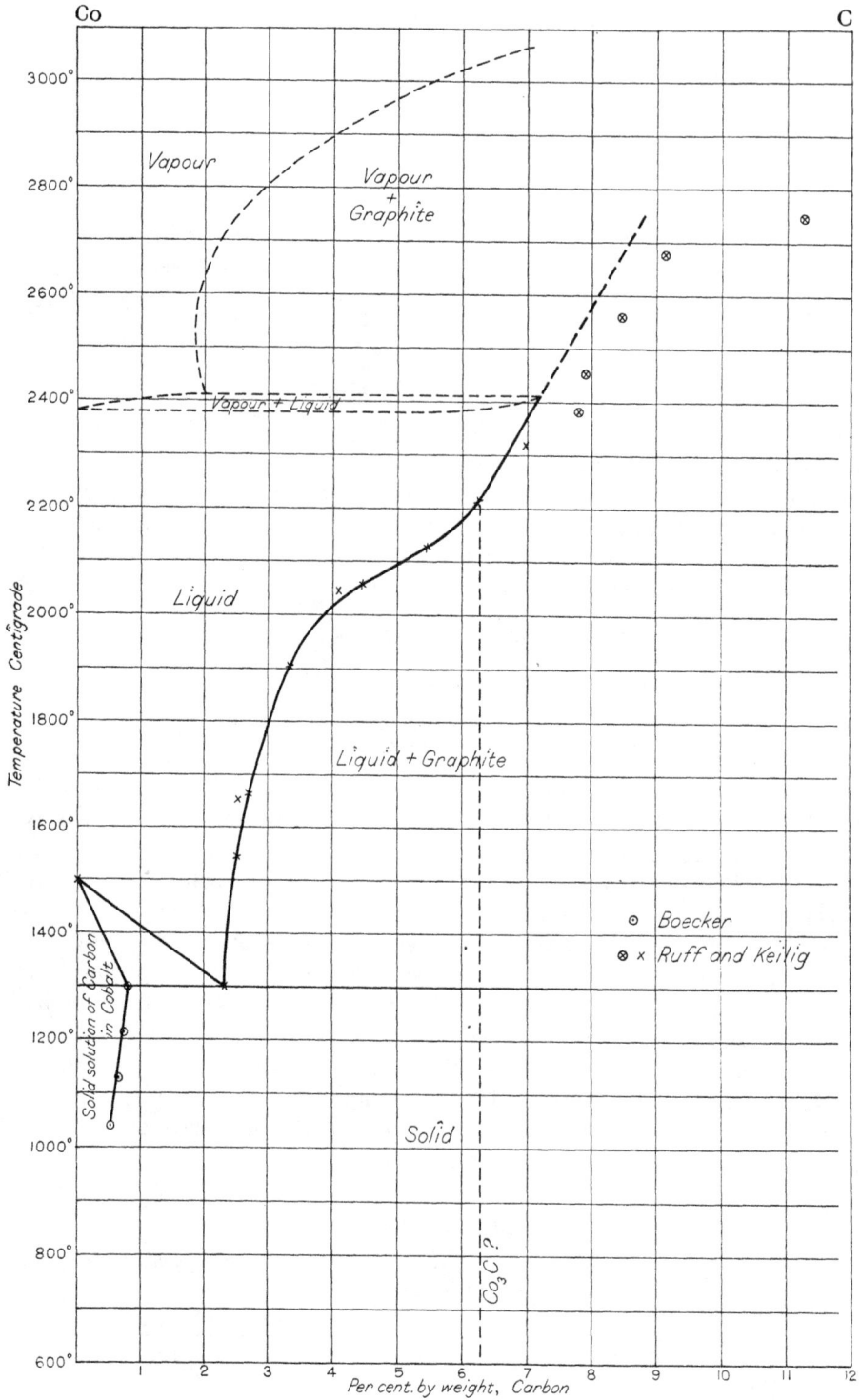

Figure 5.—Equilibrium Diagram of Cobalt-Carbon Alloys.

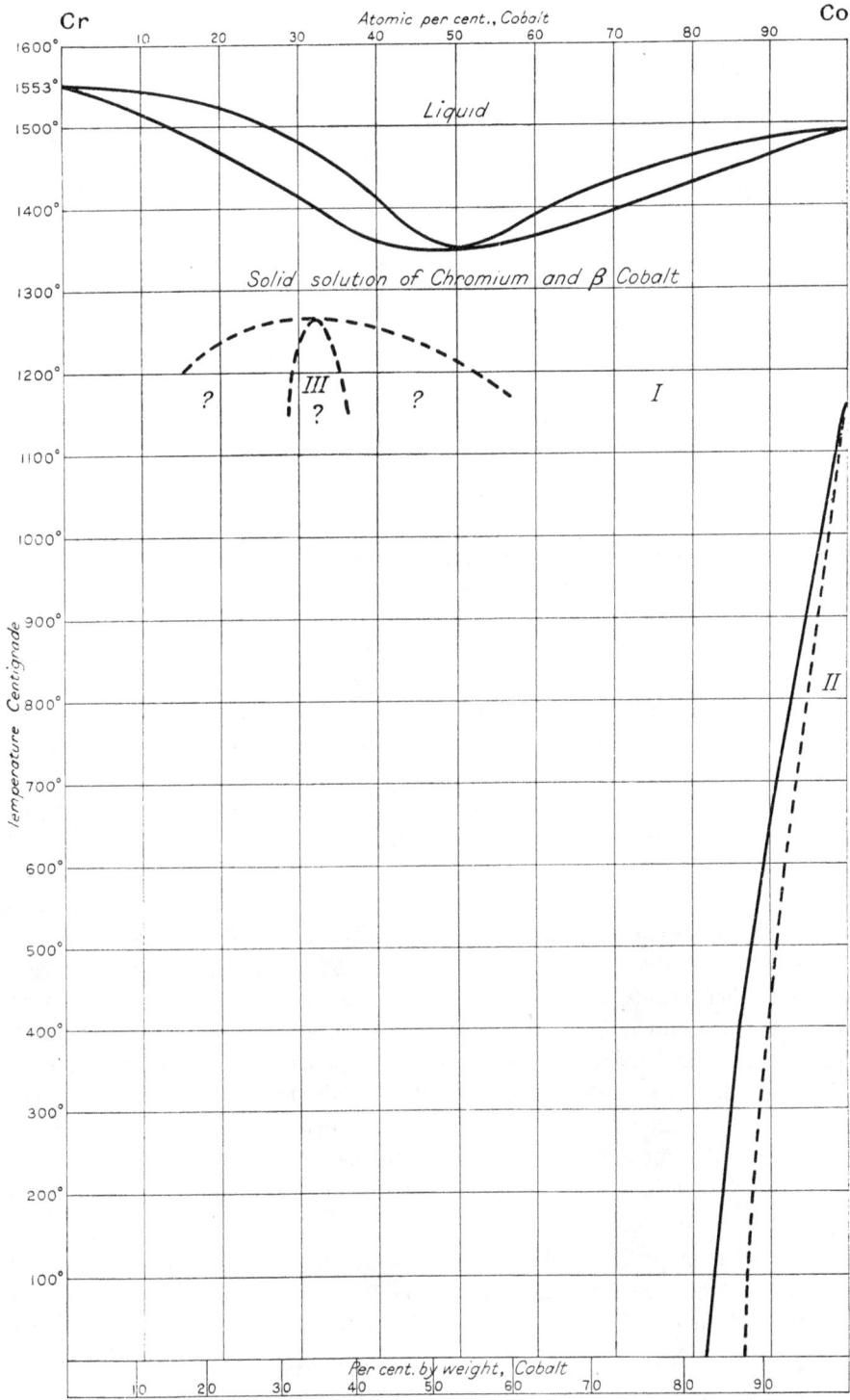

Figure 6.—Equilibrium Diagram of Chromium-Cobalt Alloys.

Cobalt and Chromium

The equilibrium diagram of the cobalt chromium alloys is shown in Figure 6.[1] Both metals are soluble in one another in the liquid and also in the solid state. The liquidus curve shows a minimum at approximately 50 per cent. cobalt and 1340°C. In alloys containing between 45 per cent. and 85 per cent. chromium there is a reaction in the solid state at approximately 1225°C., the homogeneous solid solution above that temperature breaking down into two solid solutions. Alloys with 0 to 45 per cent. chromium show a polygonal structure containing cobalt rich cores, the chromium content increasing from the centre to the outside. In alloys with more than 55 per cent. chromium, the chromium content of the grains decreases from the centre to the outside.

The temperature at which the unmagnetic cobalt chromium alloys become magnetic decreases rapidly with increasing chromium content. The addition of 10 per cent. chromium lowers the transformation point to 685°; 15 per cent. to 300°; while the addition of 25 per cent. lowers the transformation to below room temperature.

Special alloys containing cobalt and chromium are known as "stellite." The word is derived from the Latin, stella, a star, and was chosen because of the brilliant polish these alloys take and retain under atmospheric conditions. An alloy containing 75 per cent. cobalt and 25 per cent. chromium is fairly tough and hard, and may be forged. This alloy is only slightly attacked by nitric acid, and is recommended for cutlery. To increase the hardness of the stellite alloys, varying amounts of either tungsten or molybdenum, or both, are added, the chromium in the alloy remaining about 30 per cent. The addition of 5 per cent. tungsten produces a distinctly harder alloy, which forges readily. With 10 per cent. the metal may be forged, and takes a fine cutting edge. This alloy is suitable for cold-chisels and wood-working tools. With 15 per cent. the metal may still be forged, but only with great care, as it is considerably harder than the 10 per cent. tungsten alloy. With 20 per cent. the alloy is still harder. It may be forged, but only to a limited extent. This alloy is adopted for cutting steel and other metals at a moderate speed. With 25 per cent. a hard alloy is obtained which cannot be forged, but is cast into bars which are ground to a suitable form for lathe tools. These tools are used to cut steel and cast iron, and retain their hardness at high speeds. When the tungsten content reaches 40 per cent. the alloy still retains its cutting qualities, and is preferred to the 25 per cent. alloy for cast iron. Further additions of tungsten produce brittleness. It is claimed these alloys possess an advantage of 20 to 100 per cent. over high-speed tool steels.

Molybdenum produces somewhat the same effect as tungsten, only a smaller proportion is required. An alloy containing 10 per cent. of molybdenum makes an excellent lathe tool. Carbon, silicon, and boron when present impart brittleness.

---

[1] Guertler, Metallographie, Vol. I, pp. 359-361. Lewkonja, Cobalt-Chromium Alloys: Zeitschr. anorg. Chemie, Vol. LIX, 1908, pp. 323-327.

A few typical analyses[1] of stellite alloys are given below, but it must be remembered that for the successful use of stellite different alloys must be used for different classes of work.

Co	Cr	W	Mo	
75	25	..	..	Original stellite alloy.
70	25	5	..	Forges readily, suitable for wood-cutting tools and cutlery.
60	15	25	..	Suitable for lathe tools for cutting steel and cast iron.
55	35	10	..	
50	30	20	..	
55	15	25	5	High speed cutting tools.
45	15	..	40	Very hard alloy.

Stellite alloys also contain a small percentage of carbon.

The stellite alloys are covered by a number of patents, the numbers of which are given below.

Haynes, U. S. Patent No. 873,745, Dec. 17, 1907. Alloys containing 10-60 per cent. chromium, and 90-40 per cent. cobalt.

U. S. Patent No. 873,746, Dec. 17, 1907. Alloy of 30-60 per cent. chromium and 70-40 per cent. nickel.

U. S. Patent 1,057,423, and 1,057,828, April 1, 1913. The addition of tungsten and molybdenum to the cobalt-chromium alloy is specified in these patents to produce greater hardness and toughness.

U. S. Patent No. 1,150,113, Aug. 17, 1915. Alloy for tools containing approximately cobalt 25 per cent., chromium 20, and iron 55. Molybdenum may be added to vary the colour. More chromium increases the hardness and iron renders the alloy more fusible, malleable and softer.

British Patent 2,487, August 17, 1915 (similar to U. S. Patent 1,057,423).

British Patent No. 100,434, Jan. 5, 1916. Cobalt-chromium alloys containing sufficient iron or nickel to soften the metal. This alloy is known as " festal metal."

Tamman, German Patent 270,750, August 14, 1909. A cobalt-chromium alloy for machine parts, containing 20.23 per cent. chromium and 80-77 per cent. cobalt.

Alloys of nickel and copper or cobalt and one of the following metals, namely, chromium, tungsten, molybdenum, vanadium, aluminium and uranium are used extensively for thermo-electric couples and resistance elements. These are covered by a number of patents granted to Albert L. Marsh, the dates and numbers of most of them being given below.

U.S. Patents Nos. 779,090, Jan. 3, 1905; 781,288, Jan. 31, 1905: 781,289, Jan. 31, 1905; 781,290, Jan. 31, 1905; 786,577, April 4, 1905; 811,859, Feb. 6, 1906: 853,891, May 14, 1907; 859,608, July 9, 1907; 874,780, Dec. 24, 1907: 971,767, Oct. 4, 1910.

Shortly after Marsh's patents were published the General Electric Company manufactured a similar alloy, " calorite," of the composition given below. The validity of Marsh's patents has been affirmed, and at present the Hoskins Electric Company are producing nickel-chromium alloys under the Marsh patents.

---

[1] Haynes, Alloys of Nickel and Cobalt with Chromium: Jour. Ind. and Eng. Chem., Vol. II, 1910, pp. 397-401.

In the patent suit defence the General Electric Company cited Placet's patent (Br. pat. 202, 1896) as antedating the Marsh patents. It appears that Placet found that the addition of chromium to other metals increased the hardness, toughness, and electrical resistance, but was not aware of the durability of the alloy, which latter property is one of the main advantages of nickel-chromium alloys.

The Driver Harris Company manufactured a nickel-chromium alloy containing 25 per cent. iron shortly after the General Electric Company made calorite. At the present time it is understood that the Hoskins Manufacturing Company control the Marsh patents, and all nickel-chromium alloys are made under a license from them.

The analyses of the various nickel-chromium alloys are given below.

	Ni	Cr	Fe	Mn	Specific Resistance.
Calorite	65	12	15	8	110
Nichrome	60	11	25	4	105
Nichrome 2	75	11	12	2	110
Excello	85	14	0.5	0.5	92
Tophet	61	10	26	3	107
Calido	64	8	25	3	106
Chromel A.	80	20	..	2	102
Chromel B.	85	15	..	2	88
Chromel C.	60	12	25	4	106

With regard to the use of nickel-chromium alloys,[1] it may be stated that the addition of iron makes the working of the alloys much easier, but lowers the resistance of the alloy to oxidation. The great resistance to oxidation of the nickel chromium alloys is not due to the nickel or the chromium but to the scale formed. The addition of iron also lowers the resistance of the scale, especially with temperatures above 1350°F. The lower the iron content, the higher temperature at which it is possible to operate.

Nichrome " 2 " is manufactured to compete with chromel " A," but owing to the iron content cannot be operated successfully above 1700°F., whereas chromel A may be used successfully at 2000°F.

Chromel " C " was placed on the market to supply a cheap alloy that could be successfully used for small heating apparatus, e.g., electric stoves, irons, etc., where the wire is exposed to the air and not allowed to heat above 1350°F.

A report has been made that a cobalt-chromium alloy, cochrome, may be swaged into wires which are in some respects superior to nichrome wires in electric heating elements. They are less readily oxidized at high temperatures and have a higher melting point.

Additional References

Hibbard, Manufacture and Uses of Alloy Steels, Bulletin 100, 1915, p. 60.   Bureau of Mines, Washington.
Haynes, Stellite, Met. Chem. Eng., Vol. 18, 1918, pp. 541-542.
Haynes, The Development of Stellite, Iron Age, Vol. 102, 1918, pp. 886-888.
Guillet and Godfroid, Some Observations on Stellite, Revue de Métallurgie (Mémoires), Vol. 15, 1918, pp. 339-346.
Stellite, Canadian Machinery, Vol. XIX, 1918, pp. 231-235.

[1] Private communication.

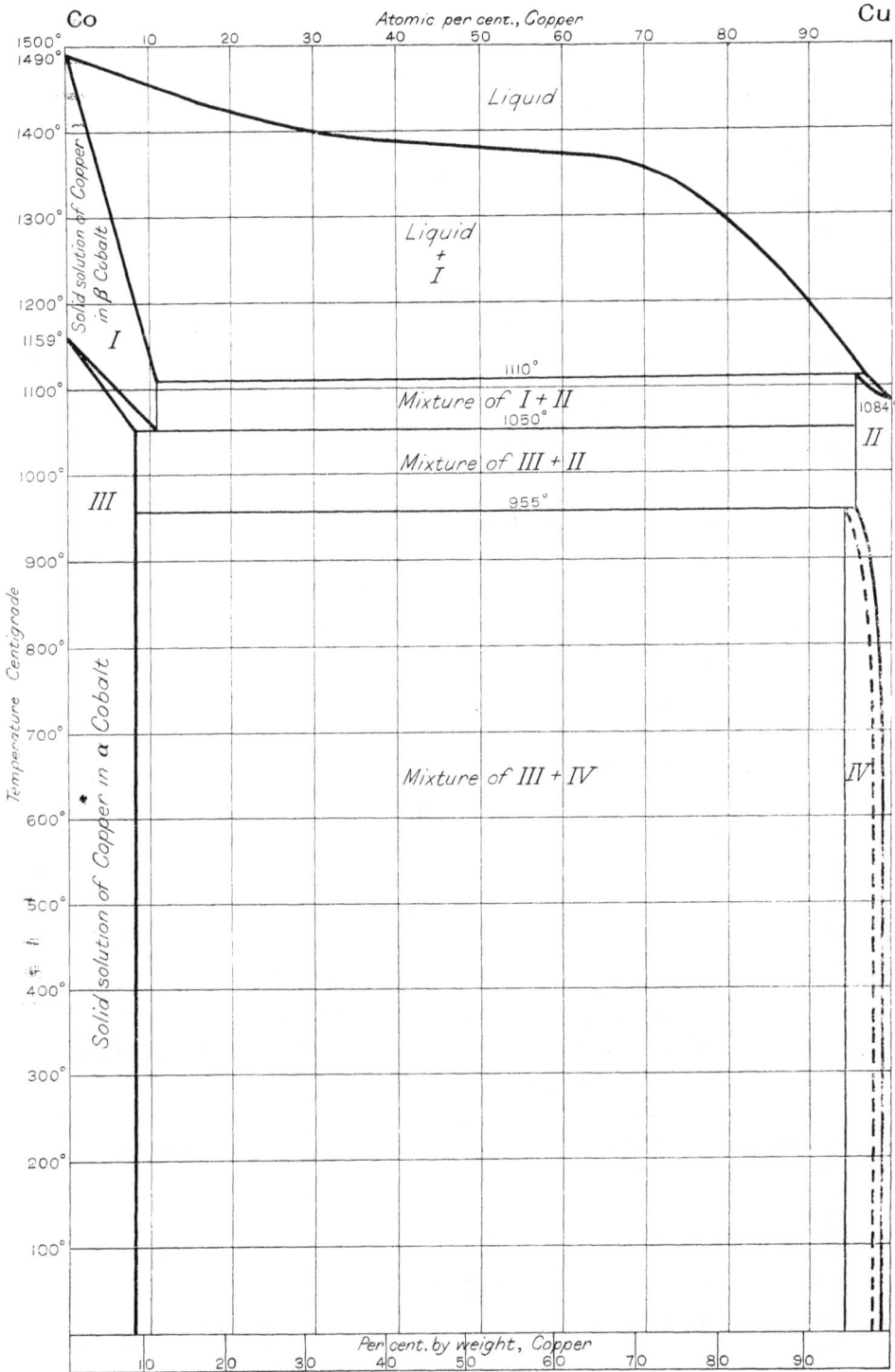

Figure 7.—Equilibrium Diagram of Cobalt-Copper Alloys.

Cobalt and Copper

The equilibrium diagram[1] of the cobalt-copper system is shown in Figure 7. The metals are completely soluble in the liquid state, but form two solid solutions in the solid. The cobalt-rich solid solution contains from 0 to 10 per cent. of copper, and the copper-rich solid solution 0 to 4 per cent. of cobalt. At 1110°C. with mixtures containing between 10 and 96 per cent. copper, there is a reaction, with the formation of the solid solution II. Below 955°C. a new solution IV containing, at room temperature, from 95 to 98 per cent. of copper, separates. This solid solution is remarkably magnetizable. The temperature of the transformation of the β cobalt solution to the α form is lowered with increase of copper from 1159° to 1050°C., where it remains constant for mixtures between 10 and 96 per cent. copper. The hardness of copper-cobalt alloys increases directly with the cobalt. Addition of either metal to the other rapidly decreases the electrical conductivity.

Additional References

Konstantinow, Alloys of Cobalt and Copper: Revue de Métallurgie (Mémoires), Vol. IV, 1907, p. 983.

Ducelliez, Studies of the Alloys of Cobalt and Copper: Procès verbaux de la société des sciences physiques et naturelles de Bordeaux, 1908-1909, pp. 120-126.

Chemical Study of Alloys of Cobalt and Copper: Bulletin de Société Chimique de France. Vol. VII, 1910, pp. 158-160, 196-198.

Waehlert, Cobalt-Nickel-Copper Alloys: Osterr. Zeitschr. Berg. u. Hüttenwesen, Vol. LXII, 1914, pp. 341, 357, 374, 392, 406.

Guertler, Metallographie, Vol. I, 1912, pp. 83-84.

Rosenhain, Electric Conductivity of Alloys of Cobalt and Copper, Introduction to Physical Metallurgy, 1914, p. 112.

Kaiser, Metallurgie, Vol. 8, 1911, p. 300. Analysis of white bronze (Cu 40-30, Co 50-60, Al 10), p. 305, Analysis of metalline (Cu 30, Co 35, Al 25, Fe 10).

Cobalt and Gold

The equilibrium diagram of the cobalt-gold series is shown in Figure 8.[2] The eutectic point which is given at approximately 90 per cent. gold and 997°C., as well as other parts of the liquidus and solidus, have not been accurately determined. In the experiments with the cobalt-rich alloys there was a marked tendency toward undercooling. Cobalt retains approximately 10 per cent. gold in solid solution, and gold retains several per cent. of cobalt, the solubility varying with the temperature. The transformation temperature of β into α cobalt has not . been determined accurately.

Additional Reference

Guertler, Metallographie, Vol. I, 1912, pp. 345-348.

---

[1] Sahmen, Alloys of Copper and Cobalt: Zeitschr. anorg. Chemie, Vol. LVII, 1908, pp. 1-9.

[2] Wahl, Cobalt-Gold Alloys: Zeitschr. anorg. Chemie, Vol. LXVI, 1910, pp. 60-72.

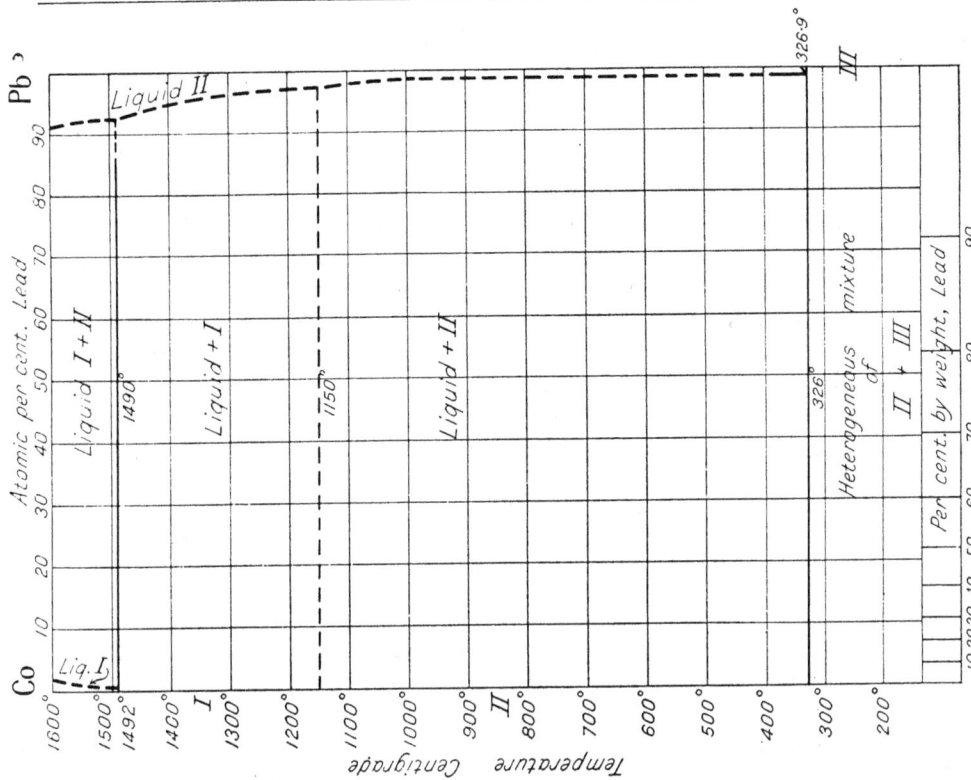

Figure 9.—Equilibrium Diagram of Cobalt-Lead Alloys.

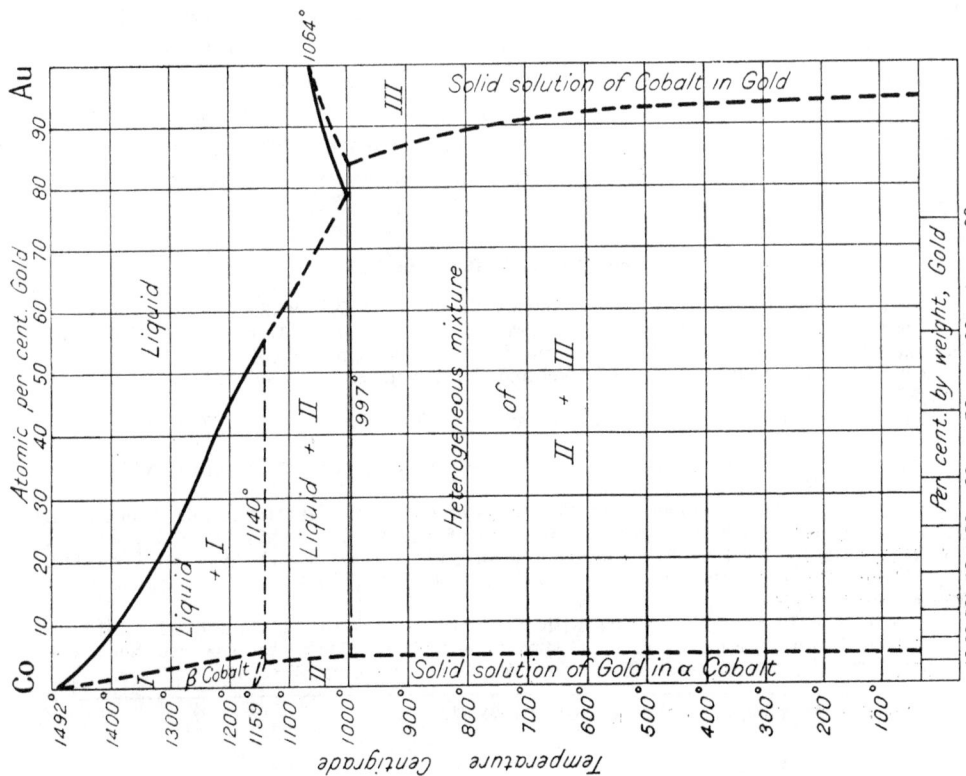

Figure 8.—Equilibrium Diagram of the Cobalt-Gold Alloys.

Cobalt and Iron

The equilibrium diagram of the cobalt-iron alloys is shown in Figure 12.[1] The two metals form a complete series of solid solutions.

A more complete diagram has been published by Ruer and Kaneko. In this diagram the two metals form a series of solid solutions with a minimum between 30 and 40 per cent. of iron and at about 1450°C. By the addition of cobalt the melting point of iron is lowered. The branch of the curve on the iron side of the diagram shows a reaction between the solid solution containing 15.5 per cent. of cobalt with the liquid containing 22 per cent. of cobalt to form a solid solution containing 19 per cent. of cobalt. The two branches of the iron curve then rise, intersecting at 1520°C., the melting point of iron.

In the same diagram the curve of magnetic transformation of iron alloys is also shown. The curve shows a maximum at 55 per cent. of iron and 985°C., which composition approaches very closely that of the compound Fe_4Co. With alloys containing less than 85 per cent. iron, the unmagnetic γ iron changes directly to the α magnetic form. With alloys containing between 100 and 85 per cent. of iron, the γ iron passes through the β form. A curve showing the transformation of β to α cobalt is also shown.

An alloy of iron, nickel, and cobalt with additions of chromium, molybdenum, tungsten, and other metals has been patented by Borchers. (For a more complete description of this alloy, see under Cobalt and Nickel.)

Yensen has investigated the magnetic properties of the iron-cobalt alloy, Fe_2Co, which contains approximately 30 per cent. of cobalt. He states that in this iron-cobalt combination, an alloy is found that is suitable for use in places where the magnetic density is very high, such as armature teeth of dynamo machinery. While its electrical resistance is low, there is reason to believe that this may be raised by the addition of other alloying elements.

Kalmus undertook a series of experiments to test whether the addition of cobalt to sheet iron had any marked effect on corrosion. The results obtained, though not conclusive, showed that cobalt in amounts up to 3 per cent. had a beneficial effect.

Additional References

Ruer and Kaneko, The Iron-Cobalt System: Ferrum, Vol. XI, 1913, p. 33.

Ruer and Klesper, The Gamma-Delta Transformation of Pure Iron, and the Effect of Carbon, Silicon, Cobalt, and Copper: Ferrum, Vol. XI, 1913, pp. 259-261.

Guertler and Tamman, Alloys of Nickel and Cobalt with Iron: Zeitschr. anorgan. Chemie, Vol. XLV, 1905, pp. 205-224.

Yensen, The Iron-Cobalt Alloy Fe_2Co, and its Magnetic Properties: General Electric Review, Vol. XVIII, 1915, pp. 881-887.

Mathews, Metallic Conduction and the Constitution of Alloys: Electrical World and Engineer, Vol. XL, 1902, pp. 531-533.

Weiss, The Magnetic Properties of the Alloys of the Ferro-Magnetic Metals, Iron-Nickel, Nickel-Cobalt, and Cobalt-Iron: mention is made of Fe_2Co. Transactions Faraday Society, Vol. VIII, 1912, pp. 149-156. Revue de Metallurgie (Memoires) Tome IX, 1912, pp. 1135-1143.

Kalmus, Magnetic Properties of Cobalt and of Fe_2Co, Bulletin 413, Department of Mines, Canada, 1916.

Kalmus and Blake, Cobalt Alloys with Non-Corrosive Properties, Bulletin 411, 1916.

Fuller, Thermo-electric Force of Certain Iron Alloys: Amer. Electrochem. Soc., Vol. XXVII, 1915, pp. 241-251; Met. Chem. Eng., Vol. XIII, 1915, pp. 318, 319.

---

[1] Guertler, Metallographie, Vol. I, 1912, pp. 78-81.

Becket, F. M., Manufacture of an alloy of approximately the composition given by Fe_2Co, containing 2 to 6 per cent. silicon. United States Patent, No. 1,247,206, Nov. 20th, 1917.

Honda and Takagi, The Irreversibility of Nickel and Cobalt Steels, Chem. Abst., Vol. XI, 1917, p. 2449.

Effect of Cobalt on Steel

Owing to the importance of nickel steels and because of the close association and similar chemical properties of nickel and cobalt, numerous experiments have been made to ascertain whether cobalt produced any marked effect similar to that of nickel on the properties of steel.

From the tests conducted so far it has been shown that cobalt has a beneficial effect on certain steels and for certain uses, but that its behaviour is very different from that of nickel.

Hadfield,[1] who was about the first to investigate the effect of the addition of cobalt to steel, concluded that cobalt raised the elastic limit and tensile strength. The amount of cobalt added varied from 0.0 to 6.9 per cent., and the steel contained 0.64 to 1.2 per cent. silicon and 0.10 to .14 per cent. sulphur.

Guillet[2] conducted a series of experiments by adding cobalt to a 0.8 per cent. carbon steel, in amounts up to 30 per cent. He also prepared a few samples containing 50 and 60 per cent. cobalt. The tests showed that the tensile strength increased with the proportion of cobalt, but that no sudden change in the mechanical properties occurred.

Arnold and Read[3] investigated the effect of cobalt on steel and concluded that the additions of cobalt increased the tensile strength, but cobalt did not show the same tendency to precipitate carbon in the form of graphite as nickel does.

Although cobalt steels have shown greater strength than the ordinary carbon steels, it is possible that this increased strength may be obtained more cheaply by the addition of other metals. However, the addition of cobalt to high-speed steels appears to produce an improvement in the cutting properties, and considerable quantities are being used at present for this purpose. In some cases it is claimed the life of high-speed steels containing cobalt, is increased several times.

The effect of cobalt on high-speed steels is quite different from that of nickel. The addition of between 5 and 10 per cent. of cobalt to steel produces a good high-speed steel, while the addition of nickel has an altogether different effect, since it produces a metal with a soft edge.

The valuable properties imparted by addition of cobalt to steel appear to be due to an increase in the hardness of the steel when at a red heat, thereby enabling the steel to cut at a higher speed.

A number of comparative cutting tests have been made by Schlesinger,[4] and the results show a distinct advantage in favour of cobalt high speed steels. One of the high-speed cobalt steels tested contained C 0.76, Si 0.28, Co 5.0, Cr 4.4, W. 16.4, and V. 0.62 per cent. This analysis corresponds very closely with an alloy steel called Rex AAA.

---

[1] Hadfield, Iron and Steel Alloys: Iron and Steel Magazine, Vol. VII, 1904, p. 10.

[2] Guillet, Cobalt Steels: Revue de Metallurgie (Memoires), Vol. II, 1905, pp. 348-349.

[3] Arnold and Read, The Chemical and Mechanical Relation of Iron, Cobalt and Carbon: Engineering, Vol. XCVIII, 1915, pp. 346-348, 362-364.

[4] Schlesinger, The Improvement of German Steel Works by the production of High-Speed Steel Alloys: Stahl und Eisen, Vol. XXXIII, 1913, pp. 929-939.

Becker[1] has obtained a patent specifying the use of cobalt in high-speed tool steels. The granting of this patent was opposed by a large number of European manufacturers, but the decision[2] was given in favour of Becker.

A later report[3] states that the patent granted to Becker has been revoked by a decision of a British Court.

Darwin and Milner[4] claim to have discovered a cobalt-chromium steel equal to the Becker iridium-cobalt-tungsten steel. One advantage of the new product is that it may be hardened at a temperature 300°C. below that required for the tungsten steel.

Additional References

Hibbard, Manufacture and Uses of Alloy Steels: Bulletin 100, 1915, p. 60, Bureau of Mines, Washington.
Chromium Nickel, and Cobalt in Pig Iron: Iron-Age, Vol. LXXX, 1907, p. 488.
Abstract of Schlesinger's tests on Cobalt Steel: Iron Age, Vol. XCII, 1913, p. 33.
Lantsberry, High-Speed Tool Steel: Iron Age, Vol. XCVI, 1915, pp. 238-241.
German High-Speed Steel: Iron Age, Vol. XCVIII, 1916, p. 1111.
Browne, Metals and Alloys in the Steel Industry: Vol. XCVII, 1916, p. 76.

Cobalt and Lead

The equilibrium diagram of the cobalt-lead series is shown in Figure 9.[5] The solubility of the two metals in the liquid state is very small.

The transformation line of cobalt was not determined, but on account of the small solubility of lead it must lie practically at 1159°C.

Additional References

Lewkonja, Cobalt-Lead Alloys: Zeitschr. anorg. Chemie, Vol. LIX, 1908, pp. 312-315.
Ducelliez, Study of the Alloys of Cobalt and Lead: Bulletin Société Chimique, Vol. III, 1908, pp. 621-622.
Chemical Study of the Alloys of Cobalt and Lead: Procès verbaux de la Société des sciences physiques et naturelles de Bordeaux, 1908, pp. 31-34.
A Study of the Electromotive Force of Cobalt-Lead Alloys: Bulletin Société Chimique, Vol. VII, 1910, pp. 201-202.
Bornemann, Cobalt and Lead: Metallurgie, Vol. VIII, 1911, pp. 361-362.

Cobalt and Magnesium

Very few experiments have been made to prepare cobalt-magnesium alloys. Parkinson[6] attempted to prepare a mixture or alloy of these metals by adding magnesium to molten cobalt. However, he obtained only small amounts of magnesium in the solid alloy. The experiments were not carried far enough to warrant any conclusions.

---

[1] German Patent, No. 281,386, Aug. 10th, 1912. Manufacture of High-Speed Cobalt Steels. An approximate analysis of a High-Speed Cobalt Steel is given as C 0.7, Cr 5, W 18, V 1, Mo 0.75 and Co 10 per cent.
 United States Patent, No. 1,081,263, December 9th, 1913. Also a British patent.
[2] Iron Age, Vol. XCIII, 1914, p. 453.
[3] Iron Age, Vol. 101, 1918, p. 321.
[4] Darwin and Milner, Cobaltchrome Tool Steel, Engineering, Vol. CIV, 1917, p. 22.
[5] Guertler, Metallographie, Vol. I, 1912, p. 582.
[6] Guertler, Metallographie, Vol. I, 1912, p. 415.

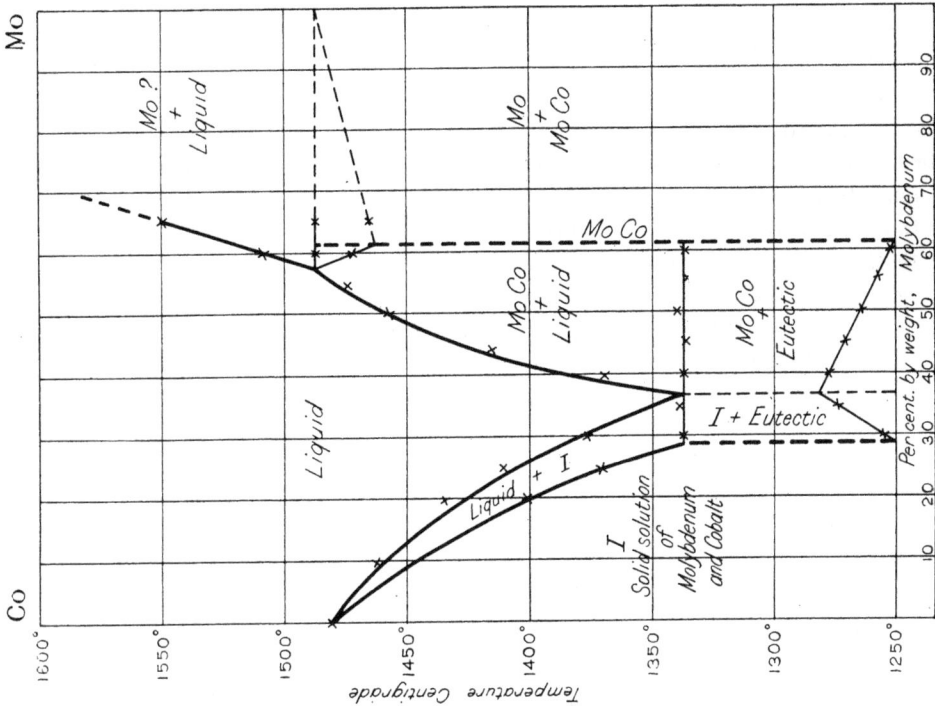

Figure 11.—Equilibrium Diagram of Cobalt-Molybdenum Alloys.

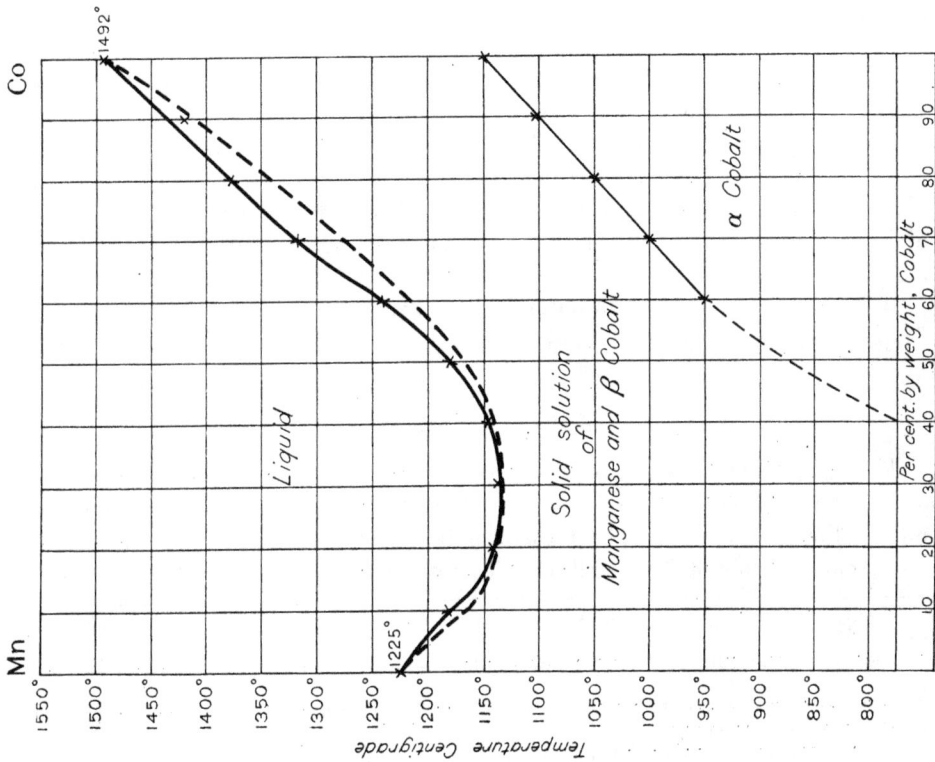

Figure 10.—Equilibrium Diagram of Manganese-Cobalt Alloys.

Cobalt and Manganese

The equilibrium diagram of the cobalt-manganese series is shown in Figure 10.[1] A minimum exists at approximately 30 per cent. cobalt and 1140°C. The alloys in solidifying exhibit a strong tendency toward undercooling. Alloys with more than 40 per cent. cobalt are not homogeneous, but become more nearly so by annealing at 1000°C. for 5 hours. Those alloys containing more than 40 per cent. cobalt were magnetic, the magnetism rising rapidly with increased cobalt content. In the original article Hiege shows the melting points of cobalt and manganese at 1525° and 1250° C. respectively. As this temperature for cobalt is about 25°C. higher than that used in the other diagrams, in reproducing the cobalt-manganese diagram all Hiege's temperatures have been reduced by 25°.

Huntington[2] states the addition of manganese to cobalt renders cobalt malleable.

Additional References

Arrivant, Alloys of Manganese with Nickel, Cobalt, and Vanadium: Procès verbaux de la Société des sciences physiques et naturelles de Bordeaux (Memoires), 1905-1906, pp. 105-114, 152-154.

Guertler, Metallographie, Vol. I, 1912, p. 90.

Guertler, The Magnetizability of the Alloys of the Ferro-Magnetic Metals, Zeitschr. physikal. Chemie, Vol. 65, 1908, pp. 73-83.

Cobalt and Molybdenum

The equilibrium diagram of the cobalt molybdenum alloys is shown in Figure 11.[3] Owing to the difficulty of obtaining homogeneous liquid melts at 1800°C., alloys with more than 65 per cent. molybdenum have not been prepared. A eutectic of a cobalt solid solution and the compound MoCo occurs at 1335°C., and 37 per cent. molybdenum. At the eutectic temperature cobalt retains 28 per cent. molybdenum in solid solution. At 1488°C., the compound MoCo (62 per cent. Mo) is formed by a reaction.

Alloys containing up to 60 per cent. molybdenum are magnetic, but those containing between 50 and 60 per cent. are very feebly so.

Additional Reference

Guertler, Metallographie, Vol. I, 1912, pp. 378-379.

Cobalt-Nickel Alloys

The equilibrium diagram of the cobalt-nickel alloys is shown in Figure 13.[4] The liquidus and solidus lie practically in a straight line between the melting points of the pure metals. The transformation temperatures of pure nickel and cobalt occur at about 320°C., and 1159°C. respectively.

[1] Hiege, Alloys of Manganese with Cobalt: Zeitschr. anorg. Chemie, Vol. LXXXIII, 1913, pp. 253-256.

[2] Huntington, Metallurgy of Nickel and Cobalt, Jour. Soc. of Chem. Ind., Vol. I, 1882, p. 258.

[3] Ravdt and Tamman, Molybdenum-Cobalt Alloys: Zeitschr. anorg. Chemie, Vol. LXXXIII, 1913, pp. 246-252.

[4] Guertler, Metallographie, Vol. I, 1912, p. 81.

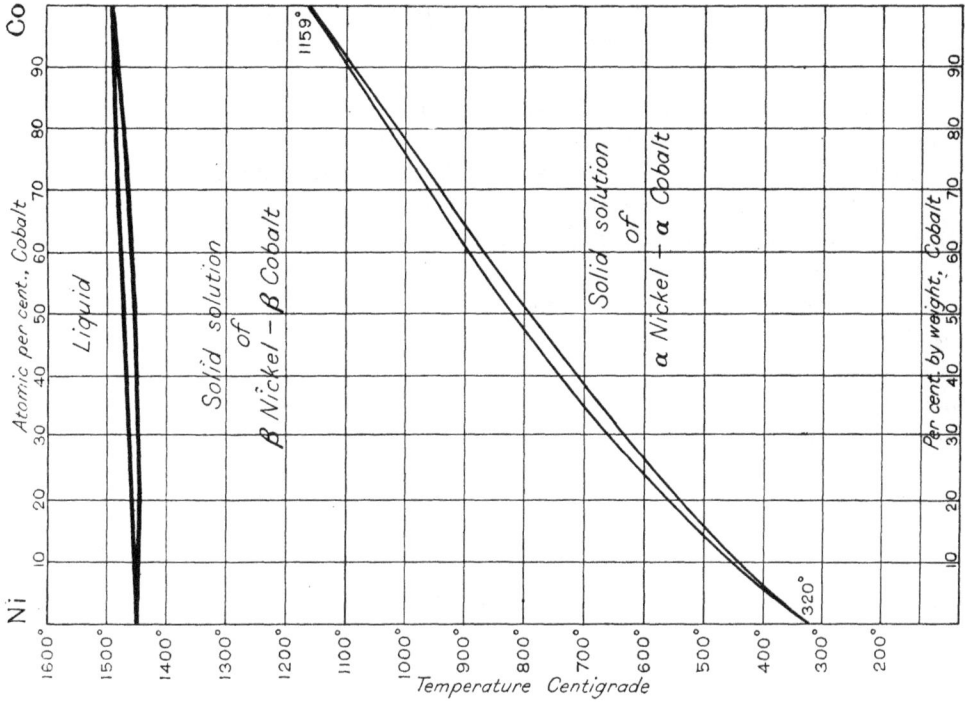

Figure 13.—Equilibrium Diagram of Nickel-Cobalt Alloys.

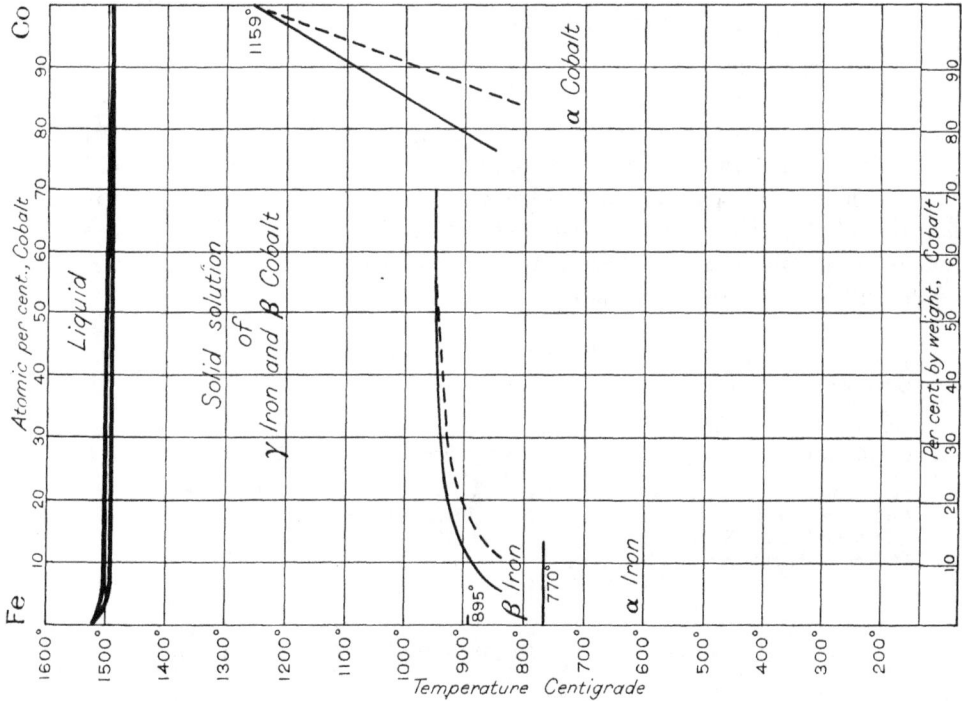

Figure 12.—Equilibrium Diagram of Iron-Cobalt Alloys.

The hardness of the cobalt-nickel alloys remains practically constant until 60 per cent. cobalt is reached, when the hardness increases rapidly with increased percentage of cobalt.

Certain alloys of cobalt and nickel are only slightly attacked by acids, and a number of patents have been granted for acid-resistant cobalt-nickel alloys with the addition of a third or fourth metal for certain other properties.

The first patent[1] was granted to Borchers for an alloy of high chemical resistance, combined with good mechanical properties. It contained approximately 67.5 per cent. nickel, 30 chromium, and 2.5 silver. The addition of silver is said to improve the mechanical properties. Later[2] copper was substituted for the silver.

The next patent[3] was for an acid resistant and mechanically workable nickel-cobalt-silver alloy, the proportions being 39.5 : 60 : 0.5 respectively. It was found later[4] that the nickel in the first patent may be replaced by an equal amount of cobalt, and the silver wholly or partly by copper. Subsequently, the silver and copper mentioned in the previous patent were replaced by molybdenum.

A later patent[5] granted for chemically resistant and mechanically workable alloys specified the substitution of gold, metals of the platinum group, or tungsten, for the molybdenum in the first mentioned patent.

Another patent[6] was granted whereby up to 90 per cent. of the nickel in the previous patents may be replaced by iron. This was followed by a patent[7] covering the alloys of iron, nickel, and cobalt and their alloys with one another, with additions of chromium between 25 and 35 per cent. and below 5 per cent. of one or more of the following metals: molybdenum, tungsten, platinum, iridium, osmium, palladium, rhodium, ruthenium, tin, silver, and copper.

Cobalt-Nickel-Copper Alloys

An extensive investigation of the cobalt-nickel-copper alloys was undertaken by Waehlert.[8] From the tests he concluded that the addition of cobalt to copper-nickel alloys shows an improvement in the tensile strength, hardness, and working properties. The ternary alloys were quite resistant to sulphuric acid (20 per cent. or more) but were attacked more vigorously by nitric acid. The greatest hardness of the series of alloys occurs when the cobalt and nickel are in approximately equal proportions. The colour of the alloys changes from copper-red to white with 50 per cent. of combined cobalt and nickel.

Additional References

Guertler and Tamman, Alloys of Cobalt and Nickel: Zeitschr. anorg. Chemie, Vol. XLII, 1904, pp. 353-363.

Weiss, The Magnetic Properties of the Alloys of the Ferro-Magnetic Metals, Iron-Nickel, Nickel-Cobalt, Cobalt-Iron: Transactions Faraday Society, Vol. VIII, 1912, pp. 149-156.

[1] German Patent, No. 255,919, June 21st, 1912, W. Borchers and R. Borchers.

[2] German Patent, No. 257,380, August 20th, 1912, Gebr. Borchers.

[3] German Patent, No. 256,123, June 21st, 1912, Gebr. Borchers.

[4] German Patent, No. 256,361, August 20th, 1912, W. Borchers and R. Borchers.

[5] German Patent, No. 265,076 and No. 265,328, Feb. 11th, 1913, W. Borchers and R. Borchers.

[6] German Patent, No. 268,516, June 12th, 1913, W. Borchers and R. Borchers.

[7] British Patent, No. 18,212, August 11th, 1913, W. Borchers and R. Borchers.

[8] Waehlert, Cobalt-Nickel-Copper Alloys: Osterr. Zeitschr. Berg. u. Hüttenwesen, Vol. LXII, 1914, pp. 341, 361, 375, and 392-406.

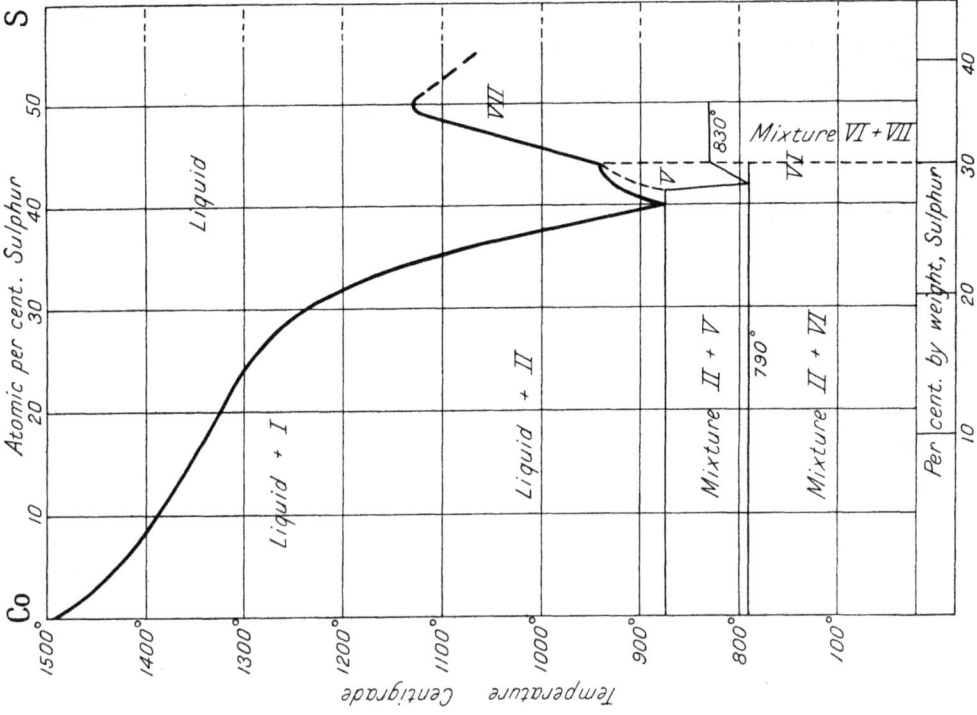

Figure 16.—Equilibrium Diagram of Cobalt-Sulphur Alloys.

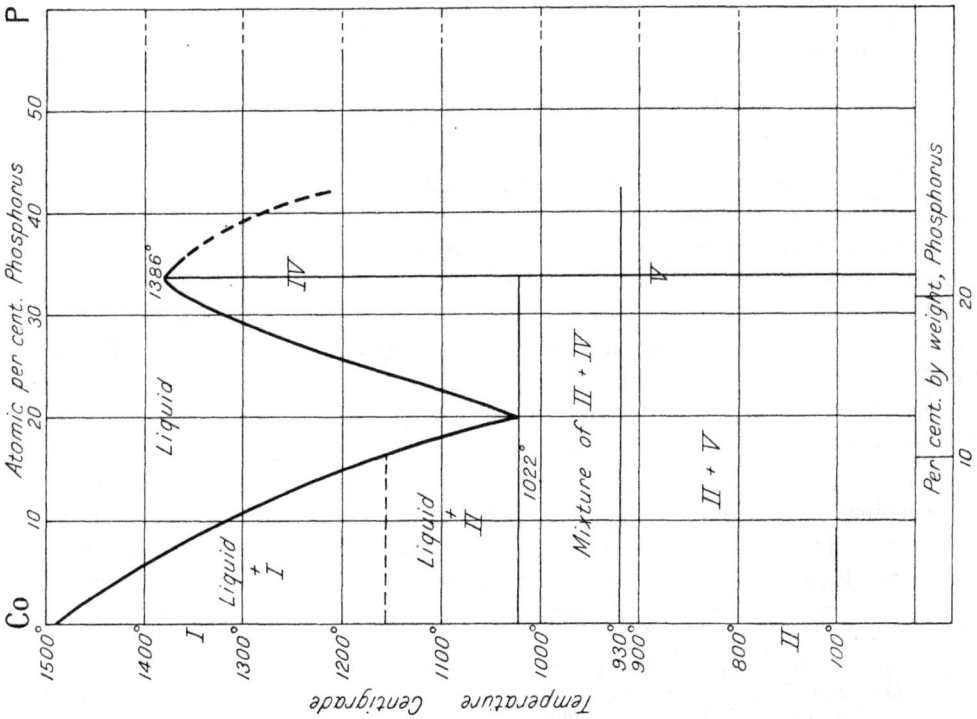

Figure 14.—Equilibrium Diagram of Cobalt-Phosphorus Alloys.

Ruer and Kaneko, The Nickel-Cobalt System: Metallurgie, Vol. IX, 1912, pp. 419-442.
    The Specific Resistance and Hardness of Nickel-Cobalt Alloys: Ferrum, Vol. X, 1913,
pp. 257-260.
    Weiss and Bloch, Magnetism of Nickel and Cobalt and the Alloys of Nickel and Cobalt:
Compt. Rend., Vol. CLIII, 1911, pp. 941-943.

Cobalt and Phosphorus

The equilibrium diagram of the cobalt-phosphorus alloys up to 20.7 per cent.
phosphorus, is shown in Figure 14.[1] The addition of phosphorus lowers the
melting point of cobalt to a eutectic point at 1022°C. and 11.5 per cent. phosphorus.
From the eutectic point the curve rises rapidly to a maximum at 1386°C., the
melting point of the compound Co_2P. Difficulty was experienced in preparing
mixtures containing higher concentrations of phosphorus.

At 920°C. the separated crystals of Co_2P undergo a transformation, forming
other crystals of similar composition.

Additional Reference

Guertler, Metallographie, Vol. I, 1912, pp. 888-890.

Cobalt and Selenium

The equilibrium diagram of the cobalt-selenium alloys has not yet been
investigated. Partial investigations seem to indicate the formation of the compounds CoSe and Co_3Se_4.[2]

Cobalt and Silicon

The equilibrium diagram of the cobalt-silicon alloys is shown in Figure 15.[3]
The curve shows two solid solutions, four eutectic points, and five compounds; one
of the compounds, mentioned again below, being formed by a reaction in the solid
state, and another by a reaction between separated crystals and liquid. Cobalt
retains 7.5 per cent. silicon in solid solution at 1204°C., while silicon retains
approximately 9 per cent. cobalt in solid solution at about 1250°C.

The eutectic point B lies at 1204°C. and 15 per cent. silicon. Further
additions of silicon cause the melting point curve to rise to C at 19.5 per cent.
silicon and 1327°C., corresponding to the compound Co_2Si. With further additions
of silicon the melting point curve is lowered to the eutectic point D at 1249°C.
and 25.5 per cent. silicon. Below 1249°C. a change occurs in the eutectic, as
shown by the line c d e, with the formation of a new compound Co_3Si_2. Further
additions of silicon cause another rise in the melting point curve to E, at approximately 1395°C., and 32.5 per cent. silicon, corresponding to the compound CoSi.
A lowering of the curve occurs again with increased silicon. At F a reaction
occurs between the compound CoSi and liquid F to form a new compound $CoSi_2$,
the composition of which is very near that of F. From F the curve drops to the
eutectic point G at 1213°C., and approximately 54 per cent. silicon. Again there
is a rise in the melting-point curve to H, at 1310°C. and 59 per cent. silicon,

---

[1] Zemczuzny and Schepelew, Phosphorus Compounds of Cobalt: Zeitschr. anorg. Chemie,
Vol. LXIV, 1908, pp. 245-257.
    [2] Guertler, Metallographie, Vol. I, 1912, p. 951.
    [3] Lewkonja, Cobalt-Silicon Alloys: Zeitschr. anorg. Chemie, Vol. LIX, 1908, pp. 317-338.
Revue de Métallurgie, Vol. 8, 1909, p. 954.

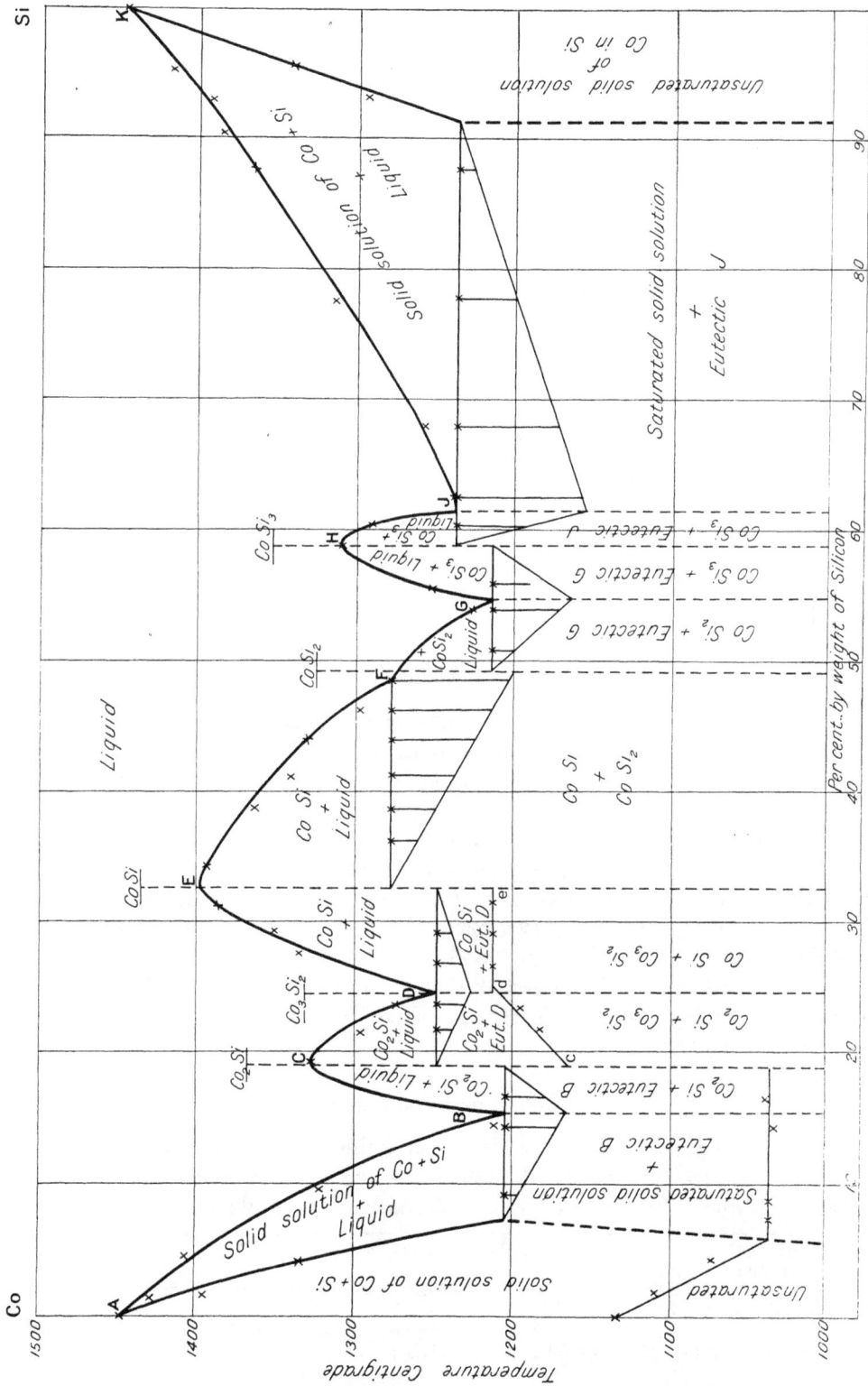

Figure 15.—Equilibrium Diagram of Cobalt-Silicon Alloys.

corresponding to the compound $CoSi_3$. The curve then falls to the eutectic J
at 1236°C., and 62 per cent. silicon. From J the melting-point curve rises to
K at 1425°C., the melting point of silicon.

Additional References

Lebeau, Silicides of Cobalt: Annales de Chimie et de Physique, Paris, Vol. XXVII, Sér. 7,
1902, pp. 271-277.
Combinations of Silicon with Cobalt and a new Silicide of this Metal: Compt. Rend.,
Vol. CXXXV, 1902, pp. 475-477.
Vigouroux, Action of Chloride of Silicon on Cobalt: Compt. Rend., Vol. CXLII, 1906,
pp. 635-637.
Beckett, F. M., Alloy containing Iron, Cobalt, and Silicon. United States Patent, No.
1,247,206, Nov. 20th, 1917.

Cobalt and Silver

The cobalt-silver alloys have been investigated by Petrenko,[1] who found
that cobalt and silver are practically insoluble in one another at temperatures
up to 1600°C.

Additional References

Ducelliez, A Study of the Alloys of Cobalt and Silver: Bulletin Société Chimique de
France, Vol. VII, 1910, pp. 506-507.
A Study of the Alloys of Cobalt and Silver: Procès verbaux de la Société des sciences
physiques et naturelles de Bordeaux, 1909-1910, pp. 46-48.
Guertler, Metallographie, Vol. I, 1912, p. 100.

Cobalt and Sulphur

The equilibrium diagram of the cobalt-sulphur alloys from 0 to 35 per cent.
sulphur, is shown in Figure 16.[2] The addition of sulphur to cobalt lowers the
melting point from 1490°C. to the eutectic at 879°C. and 73.4 per cent. cobalt.
This percentage corresponds closely to that of the formula Co_3S_2, but the eutectic
is not a compound. The curve rises with further additions of sulphur to 935°C.
and 70.7 per cent. cobalt. At this temperature the compound Co_4S_3 forms, and
from 70.7 to 73.4 per cent. cobalt, a solid solution, V, of Co_4S_3 with cobalt exists.
The liquidus curve rises finally to a maximum of 1130° and 68 per cent. cobalt,
probably approaching the compound CoS.

The formation of the compound Co_4S_3 has not yet been finally accepted.
However, there is evidence of a transformation occurring in the solid solution
V at 830° and 790°C.

Cobalt and Thallium

The equilibrium diagram for the cobalt-thallium alloys is shown in Figure
17.[3] The metals are only partly soluble in one another in the liquid state.
Thallium dissolves 2.8 per cent. cobalt, the melting point thereby being lowered

[1] Petrenko, Alloys of Silver with Iron, Nickel, and Cobalt: Zeitschr. anorg. Chemie,
Vol. LIII, 1907, pp. 212-215.
[2] Guertler, Metallographie, Vol. I, pp. 981-982. 1912. Friedrich, Diagram of Cobalt-
Sulphur Alloys: Metallurgie, Vol. V, 1908, pp. 212-215.
[3] Lewkonja, Cobalt-Thallium Alloys: Zeitschr. anorg. Chemie, Vol. LIX, 1908,
pp. 318-319. Guertler, Metallographie, Vol. I, 1912, pp. 580-582.

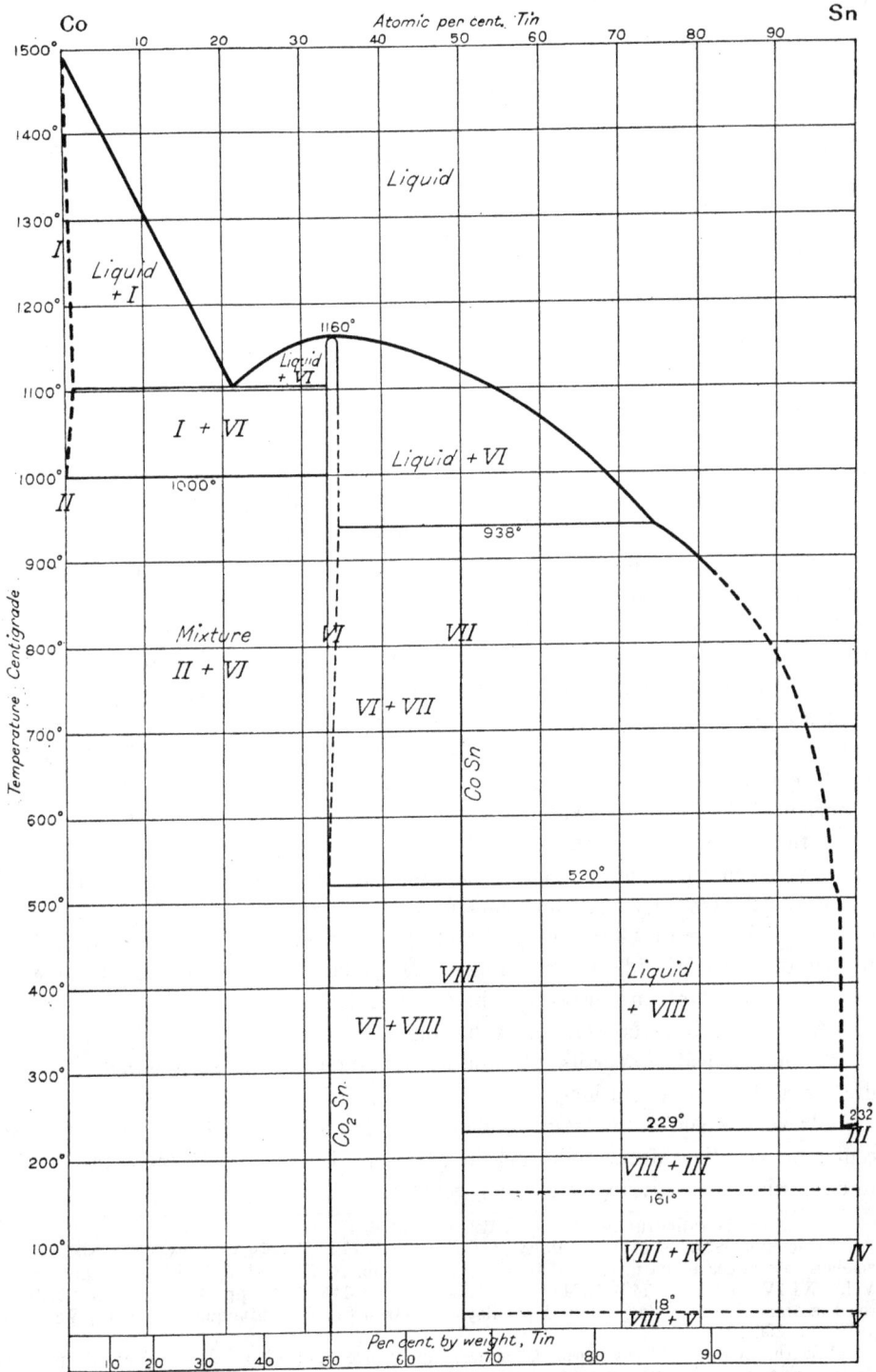

Figure 18.—Equilibrium Diagram of Cobalt-Tin Alloys.

6°C. Alloys with 10 per cent. cobalt show two layers. The boiling point of thallium lies very near the melting point of cobalt. Transformation lines are shown dotted in the diagram at 297° and 224°C.

Cobalt and Tin

The equilibrium diagram of the cobalt-tin alloys is shown in Figure 18.[1] The two metals are soluble in each other in the liquid state, but only to a small extent in the solid, cobalt retaining 2 per cent. tin at 1104°C. The addition of tin to cobalt lowers the melting point of cobalt from 1492° to the eutectic point at 1104°C. and 35 per cent. tin. With further additions of tin, the liquidus rises to a maximum at 1160°C. and 50 per cent. tin, which corresponds to the compound Co_2Sn. The curve falls with further additions of tin to 938°C., where a reaction occurs between the previously separated compound Co_2Sn and the liquid, containing 85 per cent. tin, to form a new compound $CoSn$. At 520°C., the compound $CoSn$ undergoes a transformation to form a new crystal of similar composition. At 229°C. the eutectic of $CoSn$ and tin occurs. Tin undergoes the usual transformation at 161° and 18°C.

Ducelliez[2] asserts that the compound Co_2Sn of Lewkonja is Co_3Sn_2, and has supported his assertion by numerous experiments.

An investigation of the effect of additions of cobalt on the mechanical and chemical properties of tin and bronze has been undertaken by Barth.[3] He found that adding up to 10 per cent. of cobalt to tin produced an increase in strength and an improvement in the working qualities. With 10 to 20 per cent. cobalt the alloy was a little brittle, about as hard as copper, and gray in colour. With 20 to 30 per cent. cobalt, the alloy had a smooth, bright, glassy fracture and was very hard and brittle. With 40 to 50 per cent. cobalt the hardness and brittleness increased, and the alloys broke into a number of pieces when cooling in the moulds. The alloy with 40 per cent. cobalt showed a conchoidal fracture, while that with 50 per cent. showed a very coarse crystalline structure. The alloys with more than 60 per cent. cobalt showed a finer grained, denser structure. These alloys were very hard and capable of taking a good polish.

The alloys with 20 to 50 per cent. cobalt were only slightly attacked by 60 per cent. nitric acid in five minutes. Molybdenum added to the 50 per cent. alloys did not show any advantage in the chemical tests.

A few experiments were made by Barth to investigate the effect of cobalt on the mechanical properties of bronze, but sufficient tests were not made to draw any definite conclusions.

Browne[4] made a few tests to determine the effect of cobalt (0 to 0.5 per cent.) on an alloy of the following composition: Cu 88, Sn 10, and Zn 2. The results, while encouraging, were not conclusive.

---

[1] Guertler, Metallographie, Vol. I, 1912, pp. 650-654.

[2] Ducelliez, Studies of the Alloys of Cobalt and Tin: Procès verbaux de la société des sciences physiques et naturelles de Bordeaux, 1907, pp. 51-55, 97-105, 115-119; Compt. Rend. Vol. CXLIV, 1907, pp. 1432-1434; Compt. Rend., Vol. 145, 1907, pp. 431, 502. A Study of the Electromotive Force of Cobalt-Tin Alloys: Bulletin Société Chimique de France, Vol. VII, 1910, pp. 205-206.

[3] Barth, Alloys of Increased Chemical Resistance and Good Mechanical Properties: Metallurgie, Vol. IX, 1911, pp. 261-270.

[4] Browne, Some Recent Applications of Metallic Cobalt: Trans. Amer. Inst. of Metals: Vol. VIII, 1914, pp. 61-67.

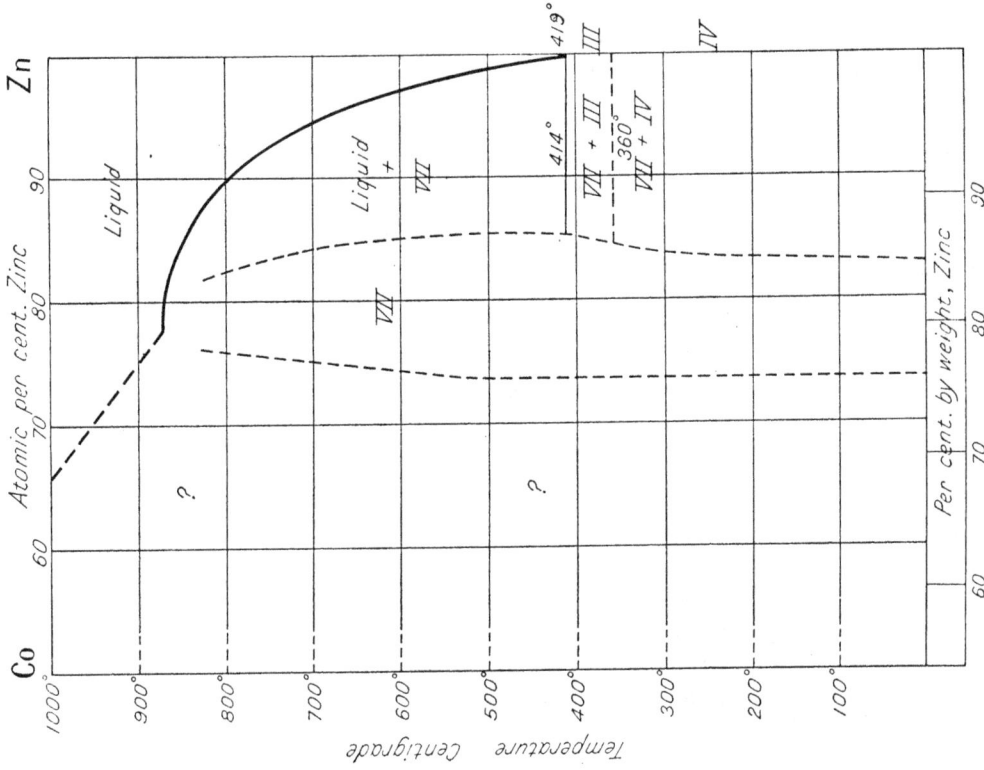

Figure 19.—Equilibrium Diagram of Cobalt-Zinc Alloys.

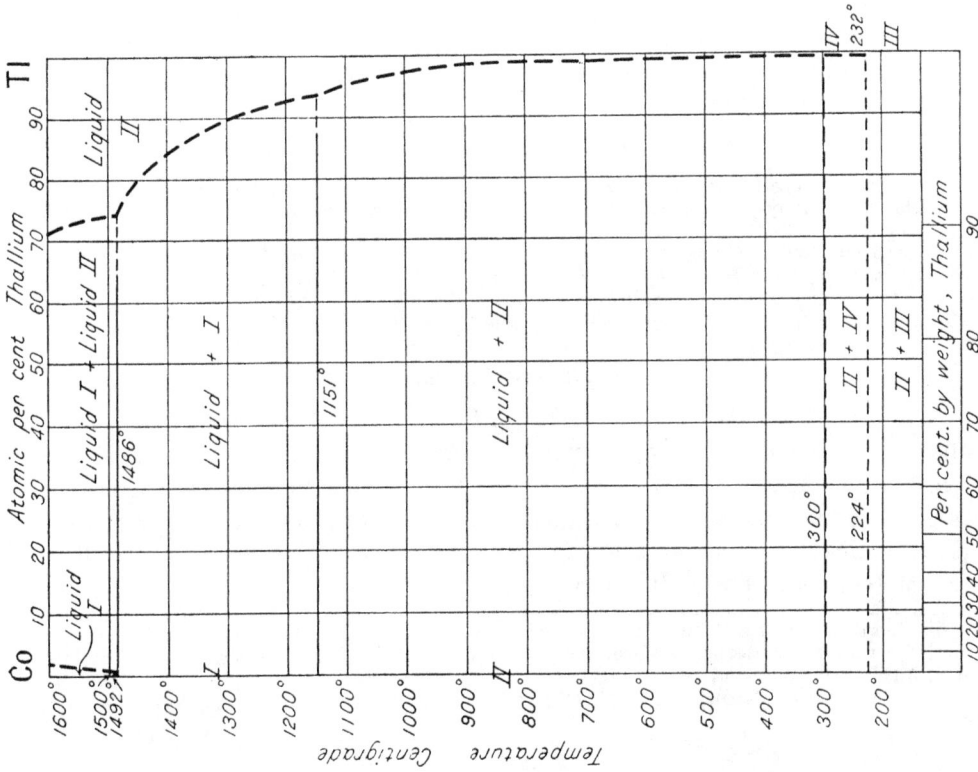

Figure 17.—Equilibrium Diagram of Cobalt-Thallium Alloys.

Additional References

Lewkonja, Cobalt-Tin Alloys: Zeitschr. anorg. Chemie, Vol. LIX, 1908, pp. 294-304.
Zemczuzny and Belynsky, Cobalt-Tin Alloys: Idem, pp. 364-370.
Bornemann, Cobalt and Tin: Metallurgie, Vol. VIII, 1911, pp. 291-292.
Kaiser, Analysis of a Tin Alloy (Sn 37, Ni 26, Bi 26, Co 11): Idem, p. 307.

Cobalt and Tungsten

The equilibrium diagram of the cobalt tungsten alloys has not yet been completely investigated. However, from the various tests that have been made, it appears that cobalt and tungsten form a complete series of solid solutions.

Reference

Guertler, Metallographie, Vol. I, 1912, p. 385.

Cobalt and Zinc

The equilibrium diagram of the cobalt-zinc alloys to 18 per cent. zinc is shown in Figure 19.[1] The addition of a small proportion (0.5 per cent.) of cobalt lowers the melting point of zinc to 414°C., 5° below the melting point. Further additions cause a rise in the liquidus, but melts containing more than 18 per cent. zinc could not be made because of the volatilization of the zinc. The curve even to 18 per cent. zinc has not been completely determined. On cooling alloys between 0.5 and 18 per cent. zinc, a solid solution varying in composition from 18.4 to 13.4 per cent. zinc separates. Melts with more than 12 per cent. zinc showed many rounded holes due to bubbles of zinc vapour.

. The alloys from 0 to 18.4 per cent. cobalt were not magnetic.

Browne[2] made a few experiments with the addition of 0.5 per cent. cobalt to manganese bronze and brass containing 80 per cent. copper and 20 per cent. zinc. Sufficient tests were not made to warrant any definite statement as to the advantage of cobalt.

Additional References

Lewkonja, Cobalt-Zinc Alloys: Zeitschr. anorg. Chemie, Vol. LIX, 1908, pp. 319-322.
Ducelliez, Alloys of Cobalt and Zinc: Bulletin Société Chimique de France, Vol. IX, 1911, pp. 1017-1023.
Chemical Study of the Cobalt-Zinc Alloys: Procès verbaux de la société des sciences physiques et naturelles de Bordeaux, 1909-1910, pp. 102-107.
Electromotive Force of Cobalt-Zinc Alloys: Idem, 1909-1910, pp. 108-109.
Preparation and Properties of the Compound $CoZn_4$: Idem, pp. 109-111.

Cobalt and Zirconium

A patent[3] has been granted to H. S. Cooper covering the preparation of an alloy of cobalt and nickel with zirconium. The inventor claims that nickel and cobalt are hardened by additions of zirconium, and that the alloy so obtained is resistant to acids and alkalies, and possesses high electrical resistance. The melting point of nickel and cobalt is lowered by additions of zirconium, but the alloys cannot be forged, drawn, or rolled.

---

[1] Guertler, Cobalt and Zinc, Metallographie: Vol. I, 1912, pp. 444, 445.
[2] Browne, Some Recent Applications of Metallic Cobalt, Trans. Amer. Inst. of Metals, Vol. VIII, 1914, pp. 61-67.
[3] United States Patent, No. 1,221,769, April 3rd, 1917. Eng. Min. Jour., Vol. 105, 1918, p. 335.

Ternary Alloys of Cobalt

Very little has been done in the scientific investigation of the ternary alloys of cobalt, except in a few special cases which are mentioned mostly under the binary alloys. Most investigations of the ternary alloys have been done mainly with the object of finding some alloy valuable industrially. Janecke[1] has made an attempt to classify the ternary alloys of the metals, Cu, Ag, Au, Cr, Mn, Fe, Co, Ni, Pd, and Pt, but the classification could not be carried to a conclusion because all the binary diagrams have not been completely determined.

Eight groups are proposed by Janecke, viz., those in which

(*a*) Each of the binary systems possesses complete miscibility in the solid state.

(*b*) One of the binaries forms a eutectic or a transition point.

(*c*) One of the binary systems possesses a limited miscibility in the liquid state.

(*d*) One of the two binary systems possesses a limited miscibility in the liquid state, and the other a eutectic or transition point.

(*e*) Two of the binaries possess eutectic or transition points, or one a eutectic and the other a transition point.

(*f*) Two of the binary systems possess limited miscibility in the liquid state.

(*g*) Of the three binary systems one possesses a limited miscibility in the liquid state, and both the others have either eutectics or a eutectic and a transition point.

(*h*) Of the three binary systems, one possesses a eutectic and two a limited miscibility in the liquid state.

Group A

The ternaries of this group may be subdivided into four groups according as the system shows (1) no minimum, (2) one minimum, (3) two minima or (4) three minima. The CoNiFe and CoFeCr systems belong to the subdivisions 3 and 4 respectively. The different binary systems are briefly summarized below.

Ternary CoNiFe.	Ternary CoFeCr.
Co-Ni, no minimum.	Co-Cr minimum (c) 1340°C. (49p.c.Co).
Co-Fe minimum (a)[2] 1500°C. (50p.c.Co).	Co-Fe minimum (d) 1500°C. (50p.c.Co).

Ni-Fe minimum (b) 1464°C. (70p.c.Ni). Cr-Fe minimum (e) 1440° C. (60p.c.Fe).

The systems CoNiFe may be represented diagramatically as in figure 20, while that of CoFeCr by figure 21.

Group B

The ternaries of this group are divided into two groups b_1 and b_2; b_1 in which one of the binaries shows a eutectic, and b_2 where one of the binaries shows a transition point. To b_1 belongs the system CrNiCo, and figure 22 represents it diagramatically. In the CrNiCo system Janecke gives a eutectic for the CrNi binary system.

---

[1] Janecke, The Ternary Alloys of the Metals, Cu, Ag, Au, Cr, Mn, Fe, Co, Ni, Pd, and Pt, Metallurgie, Vol. VII, 1910, pp. 510-523.

[2] The letters a, b, c, etc., refer to the points on the ternary diagrams.

System CrNiCo.

Cr-Ni minimum (f) 1300°C.   (42p.c.Ni).
Cr-Co minimum (g) 1340°C.   (49p.c.Co).
Ni-Co no minimum.

The system CoCuNi belongs under b_2 which is represented by figure 23.

System CoCuNi.

Co-Cu transition point (h) 1110°C.   (96p.c.Cu).
Co-Ni no minimum.
Cu-Ni no minimum.

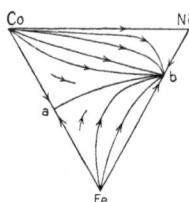

Fig. 20

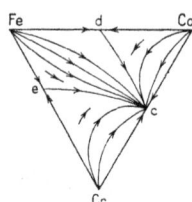

Fig. 21

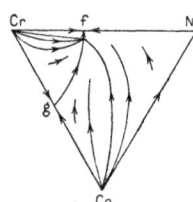

Fig. 22

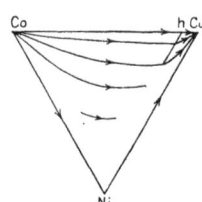

Fig. 23

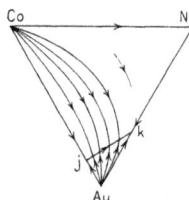

Fig. 24

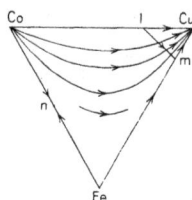

Fig. 25

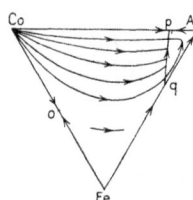

Fig. 26

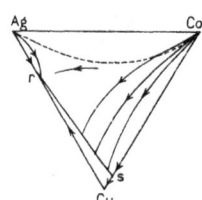

Fig. 27

Type Diagrams of some Ternary Cobalt Alloys

Group C

No system of this group is known in which all three binary mixtures have been investigated.

Group D

In Group D is given those systems in which there is in one of the binary systems a miscibility gap in the solid state, and in one of the others there occurs a gap in the liquid state. This group is subdivided according as to whether there is a eutectic (d_1) or a transition point (d_2) in one of the binary systems. The system AgCoAu belongs to d_1 while CuCrCo belongs to d_2.

System AgCoAu.	System CuCrCo.
Ag-Co two liquids.	Cu-Cr two liquids.
Ag-Au no minimum.	Cu-Co transition point.
Co-Au eutectic 997°C. (90p.c.Au).	Cr-Co minimum 1340°C. (49p.c.Co).

Group E

Group E consists of those systems in which two of the binaries show a miscibility gap in the solid state. There are three subdivisions of this group according as there are two eutectics (e_1); two transition points (e_2); or a eutectic and a transition point (e_3). To e_1 belongs CoNiAu, to e_2, CoFeCu, and to e_3, CoFeAu and AuCuCo.

System CoNiAu. (figure 24).	System CoFeCu. (figure 25).	System AuCuCo.
Co-Ni no minimum.	Co-Fe minimum (n) 1500°C. (50p.c.Co).	Au-Cu minimum 884°C. (20p.c.Cu).
Co-Au eutectic (j) 997°C. (90p.c.Au).	Co-Cu transition point (l) 1110°C. (96p.c.Cu).	Au-Co eutectic 997°C. (90p.c.Au).
Ni-Au eutectic (k) 950°C. (27p.c.Ni).	Fe-Cu transition point (m) 1100°C. (97p.c.Cu).	Cu-Co transition point 1110°C. (96p.c.Cu).

System CoFeAu. (figure 26).

Co-Fe minimum (o) 1500°C. (50p.c.Co).
Co-Au eutectic (p) 997°C. (90p.c.Au).
Fe-Au transition point (q) 1168°C. (65p.c.Au).

Group F

In this group two of the binary mixtures show a separation into two liquids, and a transition takes place in the ternary system, so that there is not only a gap in the solid state but also one in the liquid state. The mixtures NiCoAg, CoFeAg, and CoCrAg belong to this group.

System NiCoAg.	System CoFeAg.	System CoCrAg.
Ni-Co no minimum.	Co-Fe minimum 1500°C.	Co-Cr minimum 1320°C.
Ni-Ag two liquids.	Co-Ag two liquids.	Co-Ag two liquids.
Co-Ag two liquids.	Fe-Ag two liquids.	Cr-Ag two liquids.

Group G

Of the three binary systems one possesses a limited miscibility in the liquid state, and both the others have either eutectics or a eutectic and a transition point. The ternary mixture of this group may be represented by AgCoCu, figure 27.

System AgCoCu. (figure 27).

Ag-Co two liquids.
Ag-Cu eutectic (r) 778°C. (28p.c.Cu).
Co-Cu transition point (s) 1110°C. (96p.c.Cu).

9 B.M. (iii)

Group H

The binary systems of group H are the reverse of those of G. There is one eutectic, and two form two liquid layers in the binary system. No cobalt alloys are given at present under this heading.

In addition to the preceding summary by Janecke, other investigators have made experiments dealing with ternary alloys of cobalt, especially as regards their chemical and mechanical properties. A list of these is given below and a summary of the properties of the various ternaries appears under the description of the binary alloys. In order to enable ready reference to the different binaries for the summary of the ternaries, the metals are arranged in the following list so that by referring to the binary alloy of cobalt with the metal mentioned second, a description of the ternary will be found along with the references to original articles.

Cobalt-chromium-tungsten.

Cobalt-chromium-molybdenum.

Cobalt-copper-aluminium (white bronze).

Cobalt-copper-aluminium-iron (metalline).

Cobalt-iron-carbon. *See* effect of cobalt on steel.

Cobalt-nickel-silver.

Cobalt-nickel-chromium and other metals.

Cobalt-nickel-copper.

Cobalt-tin-copper.

Cobalt-tin-copper. Addition of cobalt to bronze.

Cobalt-zinc-copper. Addition of cobalt to brass.

Additional References to Alloys

Ducelliez. Researches on the Alloys of Cobalt, Thèse, Bordeaux, 1911. (Note.—A copy of this paper could not be obtained in any of the largest libraries on this continent. However, the writer believes most of the separate papers are mentioned under the alloys of cobalt with the different metals.)

Guertler, On the magnetizability of the alloys of ferro-magnetic metals: Zeitschr. physikal. Chemie, Vol. 65, 1908, pp. 73-83.

Fahrenwald, Ternary Alloys of Palladium and Gold with Cobalt, Chem. Abst., Vol. eleven, 1917, p. 1,620.

Acknowledgments

The author wishes to express his appreciation of the assistance received from Dr. William Campbell, Professor of Metallurgy, Columbia University, New York, under whom this investigation was undertaken, but besides acknowledging assistance in direction, the writer wishes to mention that the desire for research was received first from Dr. Campbell. Very few students can work under him without being impressed with his interest in investigation work, and his desire for accuracy and thoroughness in dealing with a given subject.

Acknowledgment is also made to Professor S. F. Kirkpatrick, Queen's University, Kingston, Canada, for suggestions concerning the metallurgy of cobalt, to T. W. Gibson, Deputy Minister of Mines for Ontario, Canada, for assistance in the publication of this review, and to other members of the staff of the Ontario Bureau of Mines.

END OF SECTION I.

INDEX VOL. XXVII, PART III

IMPORTANT ISOTOPES OF COBALT

Isotope	Half Life
Co-56	77.3 days
Co-57	271.8 days
Co-58	70.9 days
Co-58m	9.1 hours
Co-59	Stable
Co-60	5.3 years
Co-60m	10.5 minutes
Co-61	1.7 hours

Of these isotopes, only Cobalt-57 and Cobalt-60 have a half life longer than 80 days.

Cobalt-57 decays by electron capture.

Cobalt-60 decays by emitting a beta particle with two energetic gamma rays of 1.2 MeV and 1.3 MeV.

BIOLOGICAL ABSORPTION OF COBALT

Vitamin B_{12} is a metalloenzyme containing cobalt. Cobalt can be absorbed by inhalation or oral consumption. According to research published by Argonne National Laboratory, 50% of the cobalt which enters the body is exreted in urine almost immediately. Another 30% is exreted within 6 days. 20% more clears from the system within 60 days. The remaining 20% has a biological half-life of 800 days. This ratio does not change with age or gender.

www.ingramcontent.com/pod-product-compliance
Lightning Source LLC
Chambersburg PA
CBHW082006190326
41458CB00010B/3090